by Green Fig

How Do We Know Allah Exists?

Signs and Proofs of a Creator

DESIGN & ART BY

Guzel Murtazina

Publisher: Green Fig
Pennsylvania, USA
gogreenfig.com
info@gogreenfig.com
How Do We Know Allah Exists? -1st Edition
ISBN: 9781953836847

Introduction

﴾ سَنُرِيهِمْ آيَاتِنَا فِي الْآفَاقِ وَفِي أَنفُسِهِمْ حَتَّى يَتَبَيَّنَ لَهُمْ أَنَّهُ الْحَقُّ ﴿

It means: *"We will show them Our signs in the horizons and within themselves until it becomes clear to them that it is the Truth." (Fuṣṣilat 41:53)*

Belief in a creator is the most important certainty in one's life and the most consequential for afterlife. Our world is full of signs—around us, within us, and above us—yet many people today pass through life without noticing them. Questions about faith, purpose, and existence have become common, and some feel pressured to doubt what they once believed. This book was written to help you look again with clearer eyes and to strengthen your faith in the Creator.

Throughout history, Allah sent prophets each supported by miracles that prove the truthfulness of their message and establish their credibility. These miracles, like the fire not burning Prophet Abraham ﷺ and the cure of the one born blind and the one afflicted with leprosy by Prophet Jesus ﷺ, are facts that are passed down from generations to generations in a way that it is impossible to be a myth. The biggest miracle of all is the Qur'an which is a miracle that every living person can witness.

Beyond those miracles there are evidence and proof of a Creator in the creatures and the universe that He created. When we pause to reflect on creation—from the vastness of the universe to the intricate design within our own bodies—we discover patterns, order, and meaning that cannot be ignored. These signs strengthen the heart and open the mind. They remind us that faith is supported by reason, reflection, and evidence. The Qur'an invites us again and again to think, to ponder, and to seek certainty.

In these pages, you will explore both the signs in the world and the proofs that help us understand them. You will learn how belief rests not only on what we see, but also on the mental proofs that every sound mind can recognize. Our goal is to offer clarity and sound reasoning so that you may stand with confidence, understand your faith more deeply, and engage smartly with the questions of our time.

This journey is yours. May it strengthen your heart, sharpen your thinking, and inspire you to see the world and yourself with renewed certainty and purpose!

Contents

INTRODUCTION 3
PART 1 6
SIGNS IN CREATURES 6
CHAPTER 1 8
THE CAMEL: A SHIP OF THE DESERT 8
CHAPTER 2 12
THE BEE: INSPIRED BY ALLAH 12
CHAPTER 3 16
THE ANT: TINY BUT MIGHTY 16
CHAPTER 4 20
THE MOSQUITO: SMALL BUT STRIKING 20
CHAPTER 5 24
THE SPIDER: THE WEAKEST HOUSE 24
CHAPTER 6 28
INSPIRED TO ACT: ANIMAL BEHAVIOR AS A SIGN 28
PART 2 34
SIGNS WITHIN OURSELVES 34
CHAPTER 7 36
THE FETUS: FROM STAGE TO STAGE 36
CHAPTER 8 40
THE THUMBPRINT: A MARK OF IDENTITY 40
CHAPTER 9 44
THE HUMAN EYE: A WINDOW TO THE WORLD 44
CHAPTER 10 48
THE HEART: BEATING WITHOUT REST 48
CHAPTER 11 54
THE BRAIN: MORE COMPLEX THAN A COMPUTER 54
CHAPTER 12 58
THE DNA: A CODE WITHIN EVERY CELL 58
PART 3 62
SIGNS IN THE HEAVENS & EARTH 62
CHAPTER 13 64
THE SKY: A CANOPY ABOVE US 64
CHAPTER 14 68
THE MOUNTAINS: PEGS OF THE EARTH 68
CHAPTER 15 72
THE SUN AND THE MOON: MEASURES OF TIME 72
CHAPTER 16 76
NIGHT AND DAY: ALTERNATING SIGNS 76
CHAPTER 17 80
RAIN AND PLANTS: LIFE FROM WATER 80

CHAPTER 18 84
THE EARTH: SPREAD OUT FOR YOU 84
PART 4 88
THE MIND'S PROOFS 88
CHAPTER 19 90
THE IMPOSSIBILITY OF CIRCULAR CAUSATION (DAWR) 90
CHAPTER 20 96
THE IMPOSSIBILITY OF INFINITE REGRESS (TASALSUL) 96
CHAPTER 21 100
THE ARGUMENT FROM DESIGN AND ORDER 100
CHAPTER 22 104
THE QUR'AN: A CONTINUOUS PROOF 104
CHAPTER 23 108
THE FIṬRAH: THE NATURAL PREDISPOSITION TO ISLAM 108
CONCLUSION 112
FAITH TOOLBOX 116
THE PROOF FROM CHANGE 119

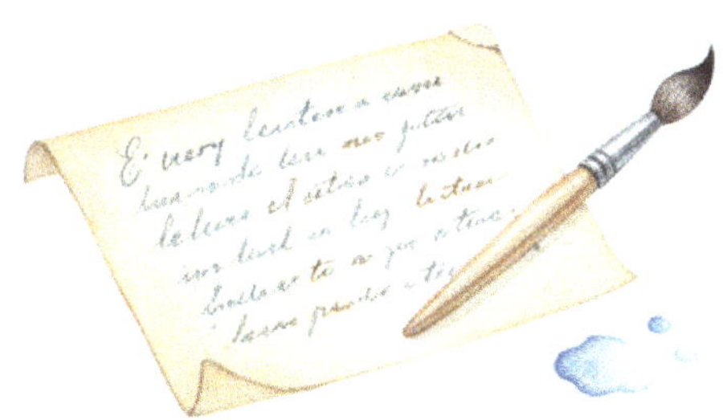

PART 1
SIGNS IN CREATURES

Every creature around us carries a sign, a message and a reminder. Animals live with abilities and instincts far beyond their size or strength. A bee builds perfect hexagons. A bird migrates thousands of miles without a map. A tiny mosquito survives on senses more precise than our machines. These are not random abilities. They are signs of guidance placed within each creature, pointing us to the One who created them and inspired them how to live.

CHAPTER 1

THE CAMEL
A SHIP OF THE DESERT

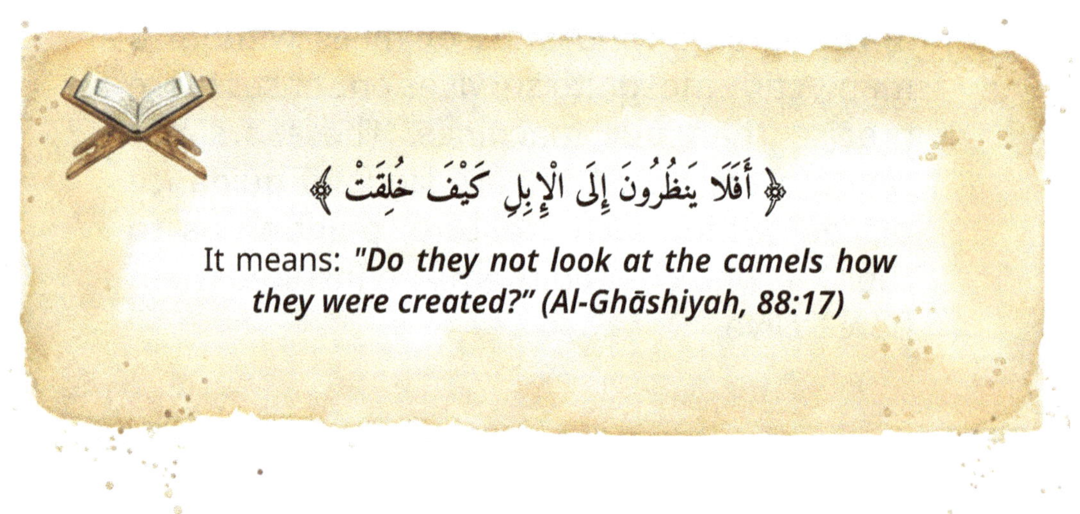

The Qur'an's Call to Reflect

Allah tells us: "Look at the camel", nor at the lion, not at the eagle. Why? Because for the Arabs who first heard the Qur'an, camels were everywhere. They were their transport, their food, their wealth and their survival. The camel was part of daily life, and yet Allah reminds us that even what you take for granted is an amazing creature when you pause to reflect. Camels are signs of Allah's power and wisdom. They have features not found in other animals. They can endure seven days without water while traveling and carry loads other livestock cannot bear. The nomadic Arabs used to place their tents on the camels' backs and move to wherever they wanted. This means that those idol-worshiping disbelievers should consider the unique characteristics of the camels.

A Creature for the Desert

The camel may look unusual, but every part of it is a sign of perfect design:

The hump stores
fat, allowing survival
when food is scarce.

Nostrils close, and long
double or triple eyelashes
interlock to protect
from blowing sand.

Wide padded feet
prevent sinking in
hot desert sands.

A thick coat keeps the
camel cool in the day
and warm at night.

A strong body allows
it to carry heavy loads
for long distances.

Water efficiency: a camel
can drink up to 40 gallons
at once and survive for
days without water.

All of this shows us that Allah created the camel perfectly suited for its environment. Nothing is accidental.

The Camel in History

For the Arabs, the camel was the ship of the desert. It carried goods across long journeys, provided milk to drink, meat to eat, wool for clothing, and strength for travel. Without camels, survival in the desert would have been very hard.

Prophet Muḥammad ﷺ rode camels, his Companions used them, and they were even part of the rites of Ḥajj. In the story of Prophet Ṣāliḥ, the miraculous she-camel was a clear sign to his people. Yet when they denied this sign and harmed it, Allah's punishment came upon them.

Lessons from the Camel

1 **Look Closer:** Even familiar things are signs of Allah's existence if we pause and reflect.

2 **Perfect Design:** Every feature of the camel is purposeful. Nothing in creation is random.

3 **Patience and Endurance:** The camel survives hardship calmly. It reminds us to be patient in our own trials.

4 **Serving Humanity:** Strong yet gentle, the camel benefits people showing us how strength is meant to serve, not just dominate.

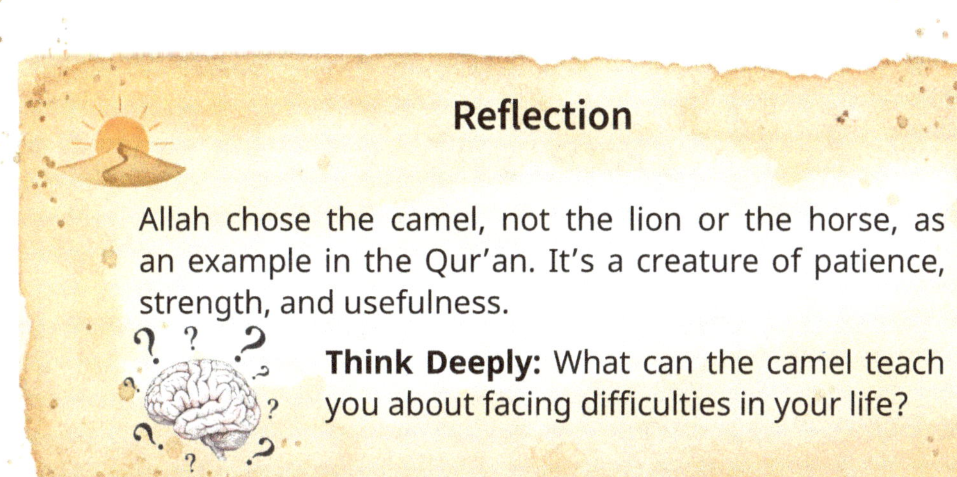

Reflection

Allah chose the camel, not the lion or the horse, as an example in the Qur'an. It's a creature of patience, strength, and usefulness.

Think Deeply: What can the camel teach you about facing difficulties in your life?

The camel is a desert wonder, created with wisdom, serving people for centuries. Allah points to it so that we learn to see His power in the ordinary and His mercy in every detail.

" O Allah, make us patient and strong like the creatures You created, and help us see Your wisdom in all things. "

CHAPTER 2

THE BEE
INSPIRED BY ALLAH

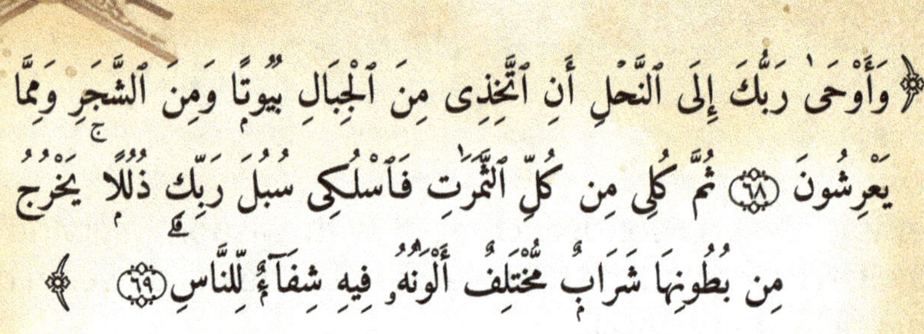

وَأَوْحَىٰ رَبُّكَ إِلَى ٱلنَّحْلِ أَنِ ٱتَّخِذِى مِنَ ٱلْجِبَالِ بُيُوتًا وَمِنَ ٱلشَّجَرِ وَمِمَّا يَعْرِشُونَ ۝ ثُمَّ كُلِى مِن كُلِّ ٱلثَّمَرَٰتِ فَٱسْلُكِى سُبُلَ رَبِّكِ ذُلُلًا يَخْرُجُ مِنۢ بُطُونِهَا شَرَابٌ مُّخْتَلِفٌ أَلْوَٰنُهُ فِيهِ شِفَآءٌ لِّلنَّاسِ ۝

It means: *"And your Lord inspired the bees: 'Take homes in the mountains, in the trees, and in what people build. Then eat from all fruits and follow the paths made easy by your Lord.' From their bellies comes a drink of varying colors, in which there is healing for mankind."*
(Al-Naḥl 16:68–69)

Allah's Inspiration to the Bee

The Qur'an uses a remarkable word: "awḥā"—Allah inspired the bee.

Usually, we think of waḥy as revelation given to prophets, but here it has the meaning of inspiration given to a tiny insect.

Wonders of the Hive

Perfect Geometry:

Bees build honeycombs of hexagons, the most efficient shape which has no gaps and no wasted space.

Social Order:

A queen bee, worker bees, drones: each has a role, and together the hive runs in harmony.

Navigation:

Bees travel long distances yet return precisely to their hives. They even communicate directions through a special "waggle dance". When a bee finds nectar, it returns to the hive and dances in a figure-eight pattern. The angle and speed of the dance tell the other bees the direction and distance of the flowers relative to the sun. It is an entire language of movement, a form of "inspiration" that scientists only recently discovered.

This kind of order and precision is far beyond human teaching. It is Allah who instilled in the bees these amazing skills.

From Nectar to Honey

Bees collect nectar from flowers, then process it inside the hive through enzymes and evaporation, turning it into honey. Honey can be golden, amber, or even dark brown, depending on the flowers. The Qur'an calls honey *"a drink of varying colors, in which there is healing for mankind."* Today, scientists know honey is antibacterial, soothing, and full of nutrients.

The Prophet ﷺ said:

عَلَيْكُمْ بِالشِّفَاءَيْنِ: الْعَسَلِ وَالْقُرْآنِ.

"Use the two cures: honey and the Qur'an." (*Ibn Mājah*).

Lessons from the Bee

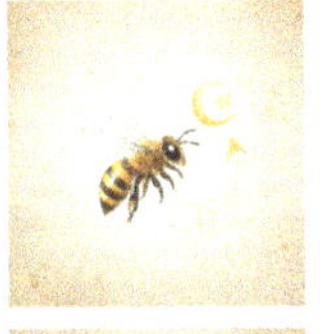 **1** **Divine Guidance:** The bee follows the guidance Allah placed within it.

 **2** **Teamwork:** No bee survives alone; their strength is in community.

 **3** **Productivity:** Every bee has a job, no idleness.

 **4** **Benefit to Others:** Bees don't only feed themselves; they provide honey for all of mankind.

Reflection

The bee is small, but it produces something sweet and healing for everyone. Its life is service and work inspired by Allah.

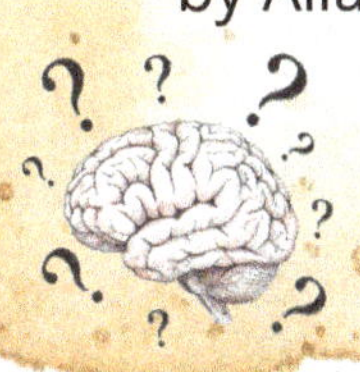 **Think Deeply:** How can you make your life like the bee, full of goodness and benefit for others?

The bee is one of Allah's tiny, amazing creatures: inspired, organized, and productive. From it, we learn that guidance comes from Allah alone, and that even the smallest creatures can bring healing to the world.

" *O Allah, inspire us to live with purpose, obedience, and benefit to others.* **"**

CHAPTER 3

THE ANT
TINY BUT MIGHTY

۞ حَتَّىٰ إِذَآ أَتَوْا عَلَىٰ وَادِ ٱلنَّمْلِ قَالَتْ نَمْلَةٌ يَٰٓأَيُّهَا ٱلنَّمْلُ ٱدْخُلُوا۟ مَسَٰكِنَكُمْ لَا يَحْطِمَنَّكُمْ سُلَيْمَٰنُ وَجُنُودُهُۥ وَهُمْ لَا يَشْعُرُونَ ۝ فَتَبَسَّمَ ضَاحِكًا مِّن قَوْلِهَا وَقَالَ رَبِّ أَوْزِعْنِىٓ أَنْ أَشْكُرَ نِعْمَتَكَ ٱلَّتِىٓ أَنْعَمْتَ عَلَىَّ... ۝

It means: *"Until, when they came upon the valley of the ants, an ant said: 'O ants, enter your homes so Sulaymān and his soldiers do not crush you while they are unaware.' So Sulaymān smiled in amusement at her words and said, 'My Lord, inspire me to be thankful for Your blessings You have given me...'" (Surah al-Naml, 27:18–19)*

A Voice from the Smallest Creature

Allah gave Prophet Sulayman ﷺ a unique gift—he could understand the speech of animals. In this verse, we learn the words of a tiny ant warning its community to take shelter so that they would not be crushed accidentally by Sulayman's army. The Qur'an shows us that even the smallest of creatures have their own form of awareness and communication. Prophet Sulayman ﷺ smiled, not in mockery, but in wonder at Allah's blessings, that a creature so small could speak with such wisdom, and that he was given the miraculous ability to understand it.

The World of Ants

Teamwork & Society:

Ants live in colonies with queens, workers, and soldiers—each with specific roles.

Communication:

They use chemical signals (pheromones) and even vibrations to pass messages.

Strength:

An ant can lift many times its own body weight. Imagine a human lifting a car!

Engineering:

Ants build complex underground homes with tunnels and chambers for food, eggs, and nurseries. Some ants return to their nest by sensing the position of the sun and counting their steps, a built-in navigation system that guides them across long distances.

Planning Ahead:

Some species seal their nests for the winter and live off the seeds and food they stored during warmer months. Others "farm" aphids for honeydew keeping them like cows to produce honeydew showing foresight and planning in their tiny societies.

The more scientists study ants, the more they discover a highly organized society—all packed into bodies just a few millimeters long.

Lessons from the Ant

1 **Wisdom Is Not About Size:** Even the smallest creature can be a teacher.

2 **Teamwork and Care:** Ants survive because they work together and protect one another.

3 **Gratitude:** Prophet Sulaymān's smile turned into a dua' of thanks showing us to connect amazement with gratitude to Allah.

4 **Allah's Power:** If Allah gave such order and wisdom to ants, how much greater is His wisdom over all creation?

A desert ant's brain is the size of a pinhead, yet it can calculate its distance and direction from its nest using the sun and a biological odometer. After wandering far in search of food, it returns straight home—an astonishing sign of the One who created it.

O Allah, inspire us to be grateful like Sulayman, and to find wisdom even in the smallest of Your creatures.

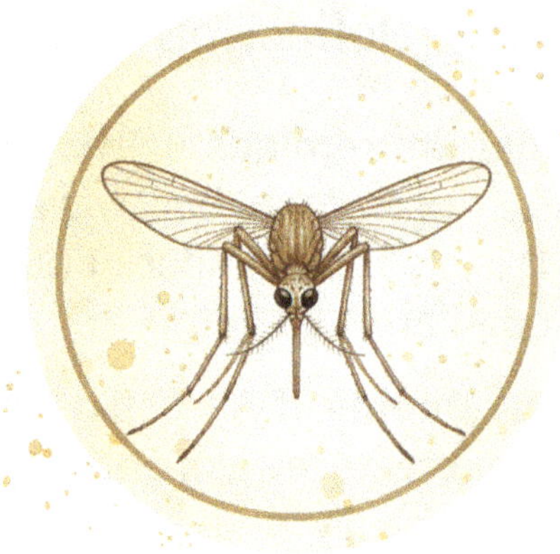

CHAPTER 4

THE MOSQUITO
SMALL BUT STRIKING

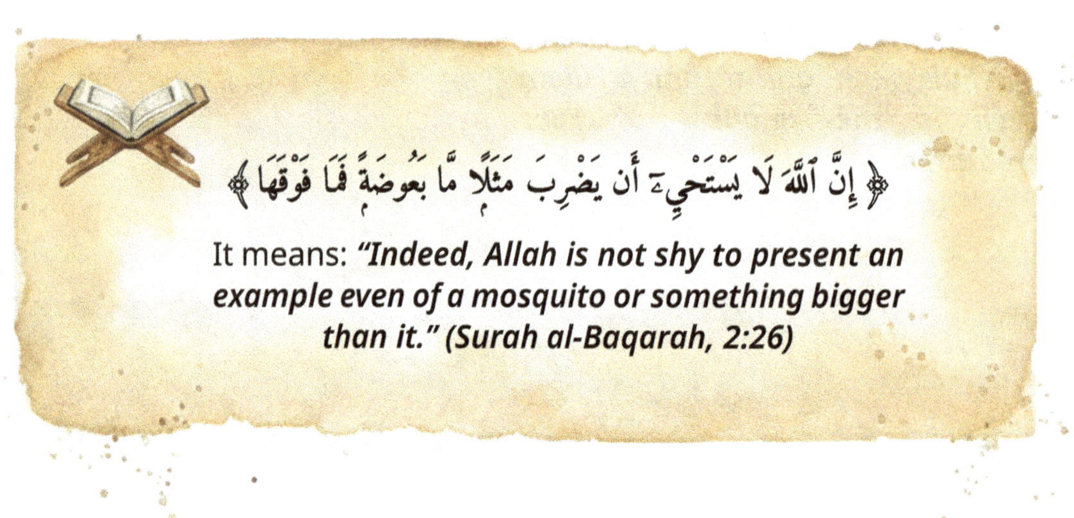

﴿ إِنَّ ٱللَّهَ لَا يَسْتَحْيِۦٓ أَن يَضْرِبَ مَثَلًا مَّا بَعُوضَةً فَمَا فَوْقَهَا ﴾

It means: *"Indeed, Allah is not shy to present an example even of a mosquito or something bigger than it."* (Surah al-Baqarah, 2:26)

Why Mention the Mosquito?

When this verse was revealed, some disbelievers mocked the Qur'an for mentioning such a tiny, irritating insect. But Allah declared that nothing is too small to be a sign. Even the mosquito, despised by many, has layers of wisdom in its creation.

The Mosquito in Science

Despite its size, the mosquito is a marvel of complexity:

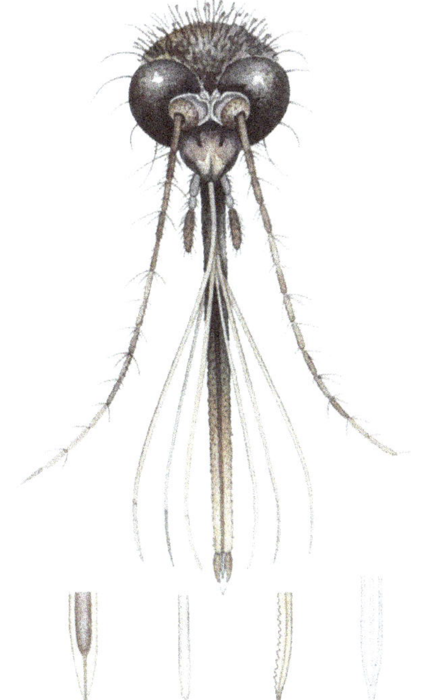

Specialized Mouthparts (Proboscis):

The female mosquito's mouth is like a surgical tool made of six tiny needles. Some pierce the skin, some search for blood vessels, and one tube draws the blood.

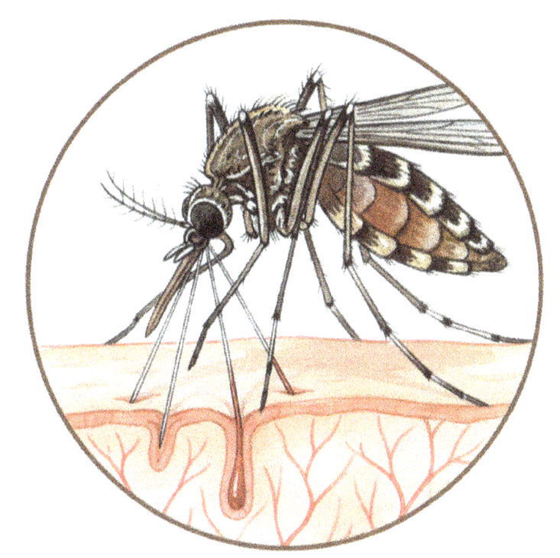

Blood Thinners:

To stop the blood from clotting, the mosquito injects saliva with anticoagulant chemicals. This is why we feel itchy afterward, our immune system reacts to those proteins.

Anesthetic Effect:

The saliva also contains substances that reduce pain so we may not even notice the bite immediately.

Targeting Blood Vessels:

Mosquitoes sense warmth and carbon dioxide from our breath to find us. They can even detect differences in blood types. Studies suggest they are more attracted to Type O blood than others.

Precision:

All of this is packed into a body smaller than a grain of rice.

Scientists continue to study mosquitoes to this day, amazed at the complex abilities built into such a tiny creature.

Lessons from the Mosquito

1 **No Creature Is Too Small:** Allah can make the tiniest insect a lesson in His wisdom.

2 **Signs of Power in Weakness:** Something so small can overpower humans, spread diseases, and change history.

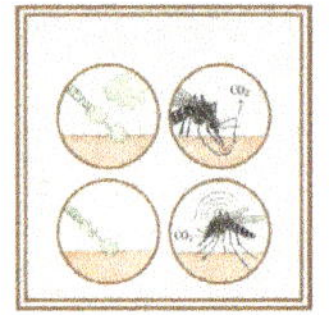

3 **Precision in Creation:** The mosquito's design—anesthetic, anticoagulant, heat sensors—shows willful creation, not chance.

4 **Reminder of Dependence:** If a small insect can cause us sickness, how much more should we remember our weakness before Allah?

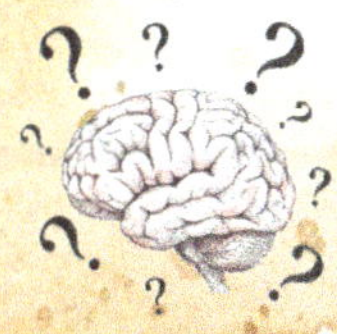

People often think signs of God's power must be in huge spectacles—mountains, planets, galaxies. But Allah tells us that even a mosquito can be a sign. Its complexity is beyond human imitation.

Think Deeply: If Allah can place such design in a creature this small, what does that teach you about His power over the whole universe?

The mosquito is tiny, often brushed away, yet it carries sophistication that scientists still struggle to fully understand. Allah chose it as an example in the Qur'an teaching us that no creation is too small to reveal His wisdom.

❝ *O Allah, let us see Your signs in the smallest and the greatest of creation, and increase our certainty in You.* ❞

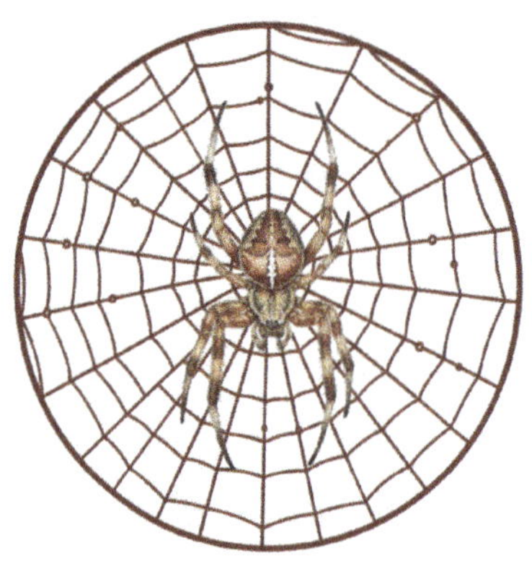

CHAPTER 5

THE SPIDER: THE WEAKEST HOUSE

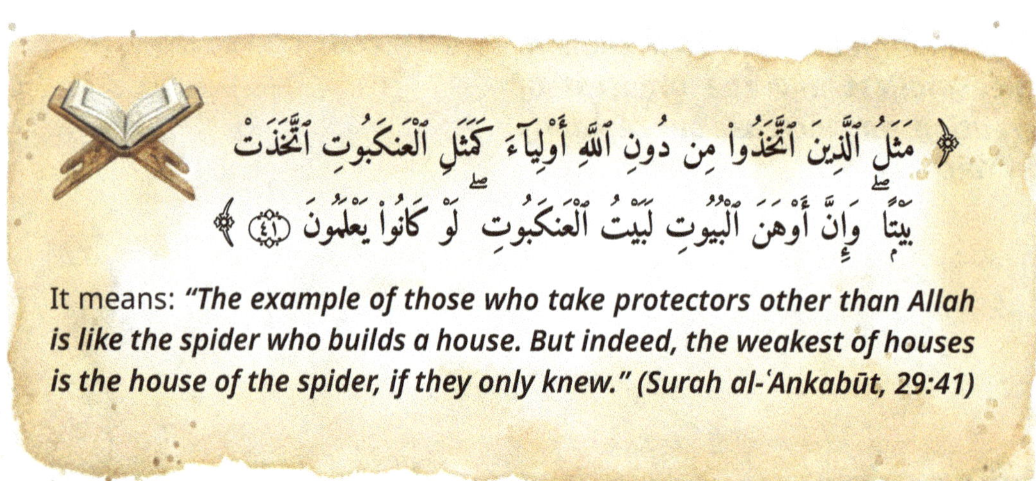

It means: *"The example of those who take protectors other than Allah is like the spider who builds a house. But indeed, the weakest of houses is the house of the spider, if they only knew." (Surah al-ʿAnkabūt, 29:41)*

The Qur'an's Example

In this verse, Allah gives a powerful image. The spider spins its web and takes it as a home. But in reality, that web cannot protect it. It is fragile, easily broken by wind, rain, or the touch of a hand. This is like the person who relies on other than Allah. Just as the spider's web is weak, anything we rely on other than Allah is weak.

The Qur'an teaches us here that true strength and safety are not found in material things, in people, or in wrongfully worshipped creatures—but only in Allah.

The Story of the Cave

The spider also appears in the Sirah. When Prophet Muḥammad ﷺ and Abu Bakr hid in the cave of Thawr during the Hijrah, their enemies came searching. Allah inspired a spider to weave its web across the cave's entrance. When the pursuers saw the unbroken web, they assumed no one could be inside. They turned away, and Allah protected His Messenger through one of the smallest creatures. This shows us that although a spider's web is weak, Allah can use even weakness as a shield when He wills.

Amazing Facts about Spiders

Spiders are fascinating creatures:

They produce silk that, for its size, has a higher tensile strength than steel.

They can spin different types of silk: sticky for webs, smooth for lining, even parachute silk for their young.

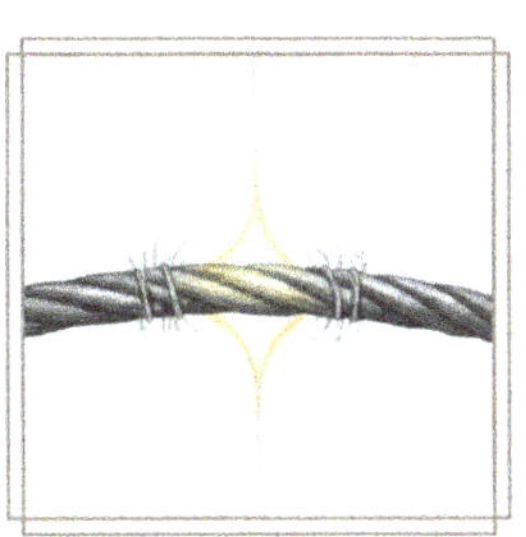

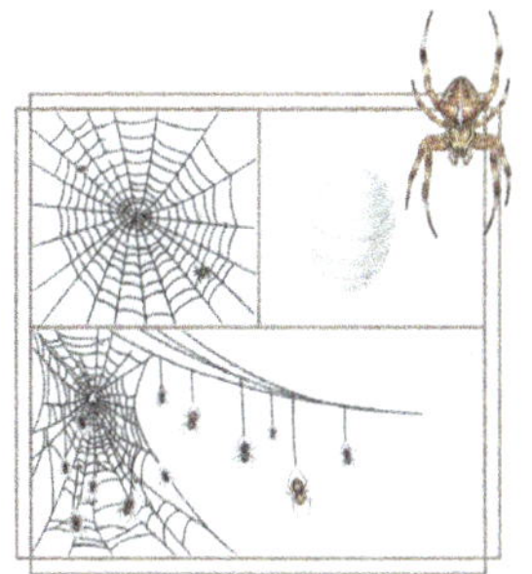

They have many eyes, helping them see in different directions.

Some spiders build intricate webs, while others hunt without webs at all.

So why does the Qur'an call their home "the weakest"? Scholars explain:

- A web cannot shield from harm; it is no real protection.

- The spider's house is also weak socially: the male often dies after mating, and sometimes the female even eats him. It is not a place of family security.

- In contrast, the home built upon faith in Allah is strong and lasting.

Lessons from the Spider

1 **Trust in Allah Alone:** Just as a web cannot truly protect, nothing in creation can protect us except by the will of Allah.

2 **Signs in Small Things:** Even fragile creatures teach us meaningful lessons.

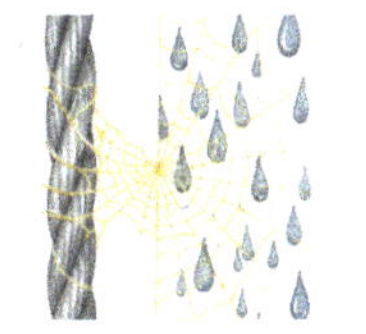

3 **Strength Is Relative:** A spider's silk is strong for catching insects but useless as a shelter. Similarly, worldly things may be useful in some ways but cannot replace trust in Allah.

4 **Allah's Wisdom:** Sometimes Allah uses even the weak (like the spider's web in the cave) to show His power and protect His beloved Prophet ﷺ.

Reflection

The spider's web looks beautiful, but it is fragile. Many things in life are like this—money, popularity, possessions—they may look enduring, but they cannot truly protect us. Only Allah can.

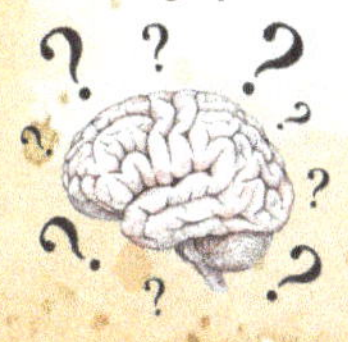

Think Deeply: What are things people today trust too much, as if they are strong, but in reality they are fragile like a spider's web?

The spider is small, its web fragile. Yet Allah chose it as a sign in the Qur'an and made it part of the story of the Prophet ﷺ. From this, we learn to never underestimate the lessons hidden in creation and never rely on anything more than we rely on Allah.

"O Allah, make our trust in You strong, and never let us cling to anything weak."

CHAPTER 6

INSPIRED TO ACT
ANIMAL BEHAVIOR AS A SIGN

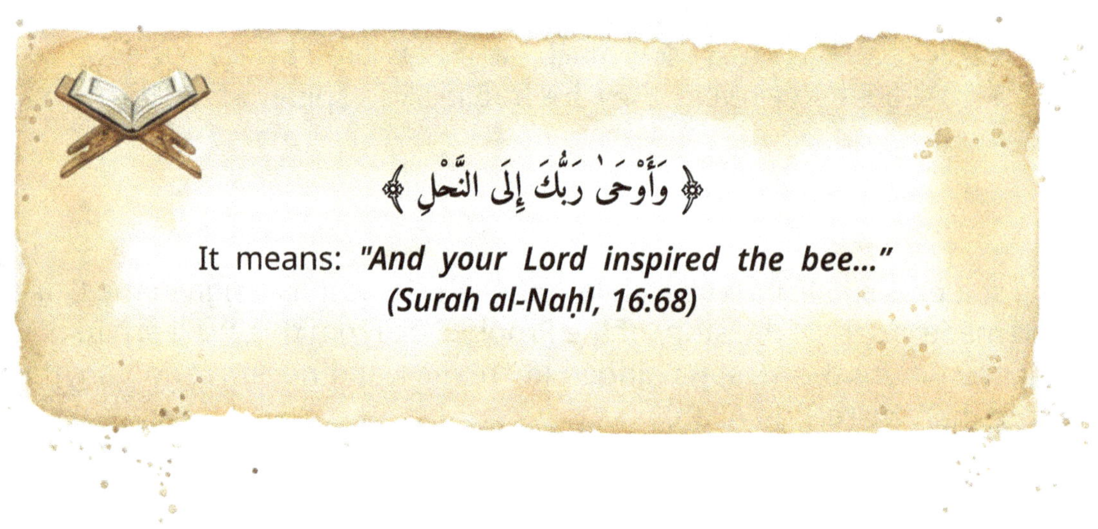

Allah Inspires Creatures

In this verse, Allah tells us that He inspired the bee to build its home and collect nectar. This shows us that creatures do not live by chance; they are guided by Allah. The bee is not alone. All around us, animals display behaviors so precise and astonishing that we cannot explain them as being an accident. They are signs of Allah's wisdom and power. All animals act with remarkable precision and instinct, not because they were taught by other creatures, but because Allah guided them. These behaviors are signs that creatures are managed by the One who knows all.

The Baby Turtle

When sea turtles hatch, they dig their way out of the sand and immediately crawl toward the ocean. No parents guide them. No one shows them the way, yet they know exactly where to go. Scientists explain that they are guided by the reflection of the moon and stars on the waves. But who gave them this guidance from birth? It is Allah who inspired them.

The Salmon's Journey

Salmon hatch in rivers, then swim out into the vast ocean. After years at sea, they return to the exact river, often to the very stream where they were born. They fight against currents, leap over obstacles, and never lose their way. Allah placed in them a sense of direction that human technology only began to imitate. This is guidance written inside them—Allah's guidance.

Birds in Migration

Every year, millions of birds travel thousands of kilometers across

continents. Some fly from the Arctic to Africa and back again, navigating mountains, oceans, and storms. They return to the same nesting grounds, generation after generation. Allah inspired them to follow these paths, providing for them along the way.

Other Examples

Monarch butterflies travel thousands of kilometers from North America to Mexico, even though no single butterfly completes the round trip journey, they pass the route on to the next generation.

Emperor Penguins gather in huge colonies, taking turns to shield one another from the

cold winds. In the freezing Antarctic, they huddle together by the thousands. They take turns moving from the outside of the group to the inside, sharing warmth in shifts. Without this cooperation, none would survive. Allah inspired them to sacrifice comfort for the survival of the group.

The Weaver Bird makes nests that look like woven baskets, hanging from tree branches. They use grass and twigs, tying knots tighter than human fingers could. No school teaches them architecture. Allah inspired them with this skill.

The Arctic Tern is a tiny bird that migrates farther than any other creature—from one end of the world to the other. It travels over 70,000 km (43,000 miles) from the Arctic all the way to Antarctica and back every year. Despite storms and oceans, it returns to its nesting place with accuracy. Allah shows us signs of His power in their flight.

Lessons from Inspired Animals

1 **Guidance Comes from Allah:** These creatures act with precision though they were never taught by other creatures.

2 **Precision without Training:** From the turtle to the salmon, their journeys are inspired by Allah alone.

3 **Strength in the Small:** From turtles to butterflies, even fragile creatures achieve the amazing by Allah's command.

4 **Dependence on the Creator:** Their instincts remind us that everything is dependent on Allah—and so are we.

Reflection

These animals remind us that guidance is not only about intellect. Allah guides every creature in its own way. We often think we are the ones in control. But these creatures remind us: all guidance comes from Allah.

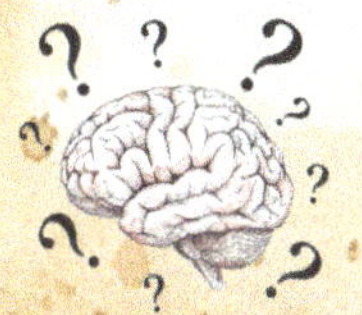

Think Deeply: If Allah inspires turtles, birds, and fish to follow their paths, how much more will He guide you if you rely on Him?

The turtle crawling to the sea, the salmon leaping upriver, the bird flying across continents—all of them are testimonies of Allah's wisdom in creation. Their behavior is not random; it is a sign for those who reflect.

> **" O Allah, guide our hearts to You just as You guide every creature in its path. ,,**

CLASSIFICATION OF ANIMALS BY HABITAT AND SIZE

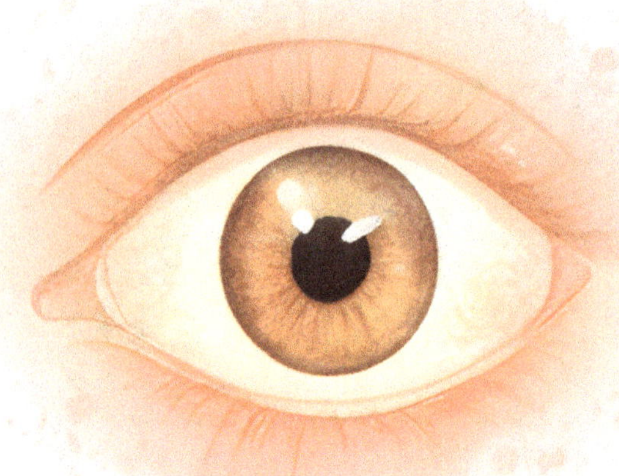

PART 2
SIGNS WITHIN OURSELVES

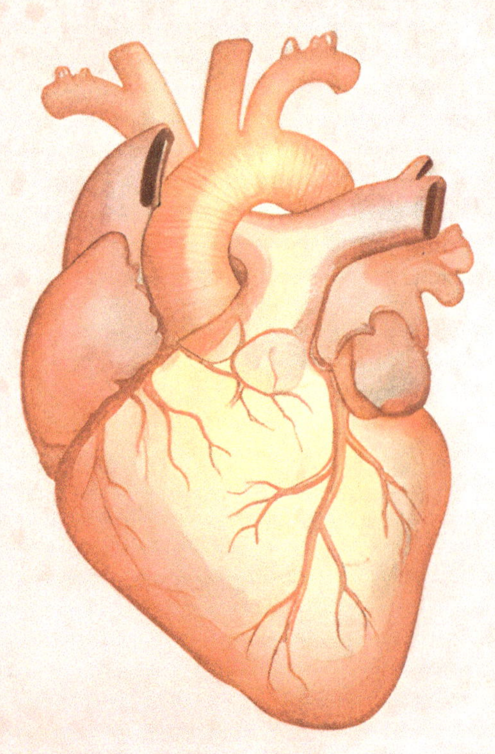

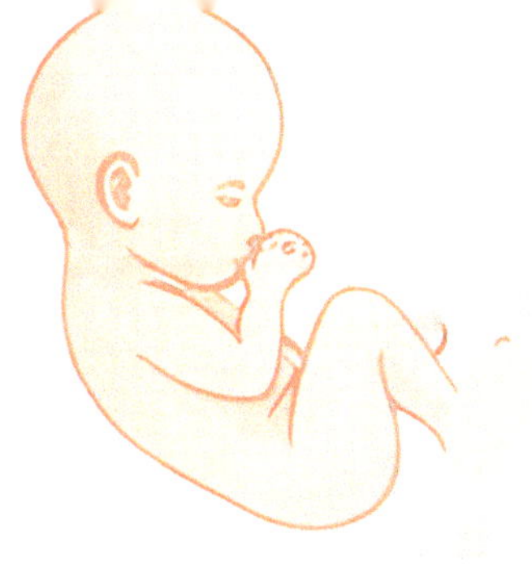

In our initial creation and the transitions from one state to another, both internally and externally, there are wonders and marvels of creation that bewilder the mind. Consider the hearts and what is embedded in them in intellect, the tongues, speech, and the mechanisms of articulation and pronunciation, with their composition and order, and the subtleties that indicate the wisdom of their Creator. The same applies to the ears, eyes, limbs, and other organs and their purposes. The joints in the limbs allow for bending and flexibility; if something stiffens, it leads to disability, and if it becomes too loose, it results in weakness. Praised is Allah.

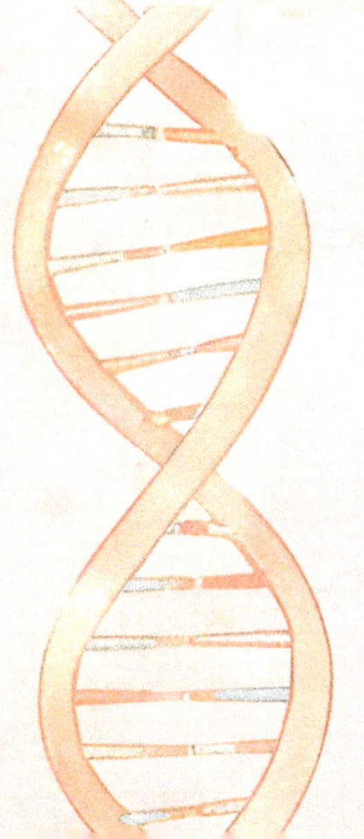

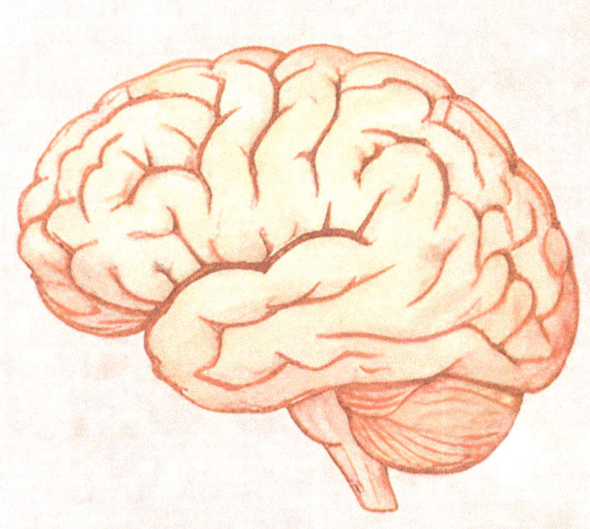

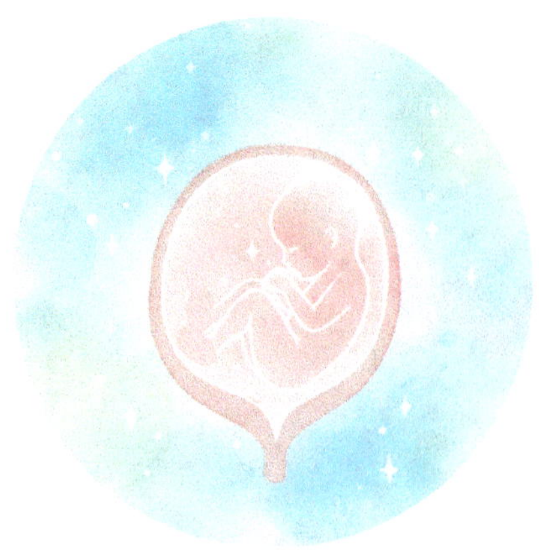

CHAPTER 7

THE FETUS: FROM STAGE TO STAGE

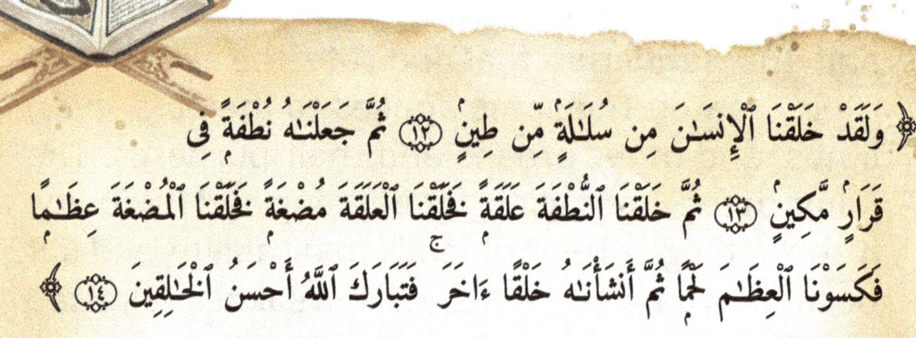

وَلَقَدْ خَلَقْنَا ٱلْإِنسَٰنَ مِن سُلَٰلَةٍ مِّن طِينٍ ۝ ثُمَّ جَعَلْنَٰهُ نُطْفَةً فِى قَرَارٍ مَّكِينٍ ۝ ثُمَّ خَلَقْنَا ٱلنُّطْفَةَ عَلَقَةً فَخَلَقْنَا ٱلْعَلَقَةَ مُضْغَةً فَخَلَقْنَا ٱلْمُضْغَةَ عِظَٰمًا فَكَسَوْنَا ٱلْعِظَٰمَ لَحْمًا ثُمَّ أَنشَأْنَٰهُ خَلْقًا ءَاخَرَ فَتَبَارَكَ ٱللَّهُ أَحْسَنُ ٱلْخَٰلِقِينَ ۝

It means: *"And indeed We created man from an extract of clay. Then We placed him as a drop in a secure resting place. Then We made the drop into a clinging clot, then We made the clot into a lump of flesh, then We made the lump bones, then We clothed the bones with flesh, then We produced him as another creation. So glorified is Allah, the best of those who decree." (Al-Mu'minūn, 23:12–14)*

From a Drop to a Human

The Qur'an describes the development of a human being in stages, beginning from a tiny drop and ending with a fully formed child. Today, scientists study embryology with microscopes, ultrasounds, and advanced technology, but the Qur'an described these stages more than 1,400 years ago.

Nutfah (Drop of fluid):

Fertilization begins when the male and female fluids combine, forming a cell invisible to the naked eye.

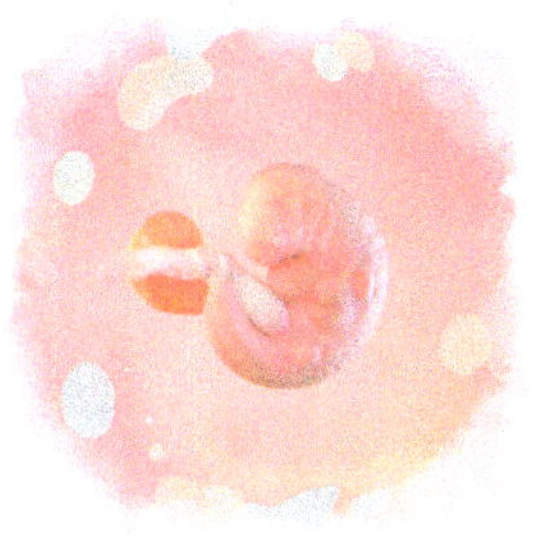

ʿAlaqah (Clinging clot):

The embryo attaches to the wall of the womb and begins to draw sustenance.

Mudghah (Chewed-like lump):

The embryo takes a shape with ridges resembling teeth marks—an image remarkably similar to what scientists see in early development.

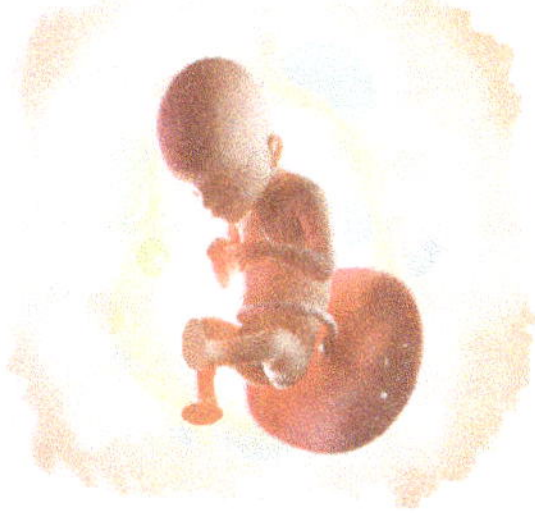

Bones, then flesh:

The skeleton forms first, followed by muscles and flesh that cover it.

Another creation:

The soul (rūh) is breathed into the child at 120 days from conception—making it not just a body, but a living being.

Scientific Astonishment

- The embryo grows at an astonishing pace from a tiny cell to a fully formed baby in just nine months.

- In the womb, the baby receives nutrients, oxygen, and protection through the placenta—system perfectly designed.

- Each stage is precisely timed. If the order were disrupted, the fetus would not form.

- Even with modern science, doctors are still amazed at the complexity of fetal development.

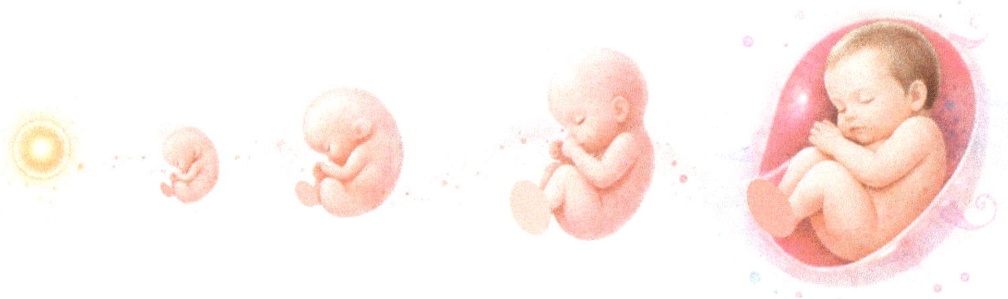

Lessons from the Fetus

1 **Human Weakness:** We began as something small and fragile, a reminder of our dependence on Allah.

2 **Stages and Growth:** Just as growth in the womb happens in stages, our spiritual growth also has stages.

3 **Proof of Resurrection:** The One who creates us in the darkness of the womb can certainly bring us back to life after death.

4 **Gratitude:** Every stage is a mercy. None of it is in our control. Our existence itself is a blessing to be thankful for.

The creation of the fetus is one of the clearest signs that Allah is the Creator. From a single drop to a complete human being, every step unfolds with wisdom.

> **❝** *O Allah, You created us stage by stage in our mother's wombs. Keep guiding us stage by stage through our lives, until we return to You.* **❞**

CHAPTER 8

THE THUMBPRINT
A MARK OF IDENTITY

It means: *"Yes, We are able to perfectly reconstruct his very fingertips."* (Surah al-Qiyāmah, 75:4)

A Unique Mark on Every Human

The Qur'an draws our attention to the banān—the fingertips. For centuries, people may not have thought much about them. But today, science shows us something extraordinary; every human being has a unique fingerprint pattern, different from all others, even if the population reaches billions.

The Science of Fingerprints

Non-repeatable Identity:

No two fingerprints are alike, not even among identical twins.

Formed in the Womb:

Fingerprint patterns develop before birth, around the 10th week of pregnancy, shaped by tiny changes in the womb environment. Once formed, they never change.

Permanent Record:

Fingerprints remain the same throughout life. Even if the skin is damaged, the pattern grows back in the same way.

Used in Technology:

From unlocking phones to border security, the uniqueness of fingerprints has become a foundation of modern life.

Why the Qur'an Mentions Fingertips

When the Qur'an says Allah can reconstruct the very fingertips, it emphasizes not just that Allah can resurrect humans, but that He can return each person with their exact, unique identity. The fingertip is a powerful symbol of individuality that sets every person apart.

Lessons from the Thumbprint

1 **Every Human Is Unique:** Out of billions, no one has your fingerprint. This shows Allah's perfect knowledge and power to create.

2 **Proof of Design:** Such complex uniqueness cannot be the product of blind chance.

3 **A Reminder of Accountability:** If Allah gave us unique identities, He will also hold each of us individually accountable.

4 **The Qur'an's Miracle:** Long before science discovered fingerprints, the Qur'an drew attention to the fingertip as a sign.

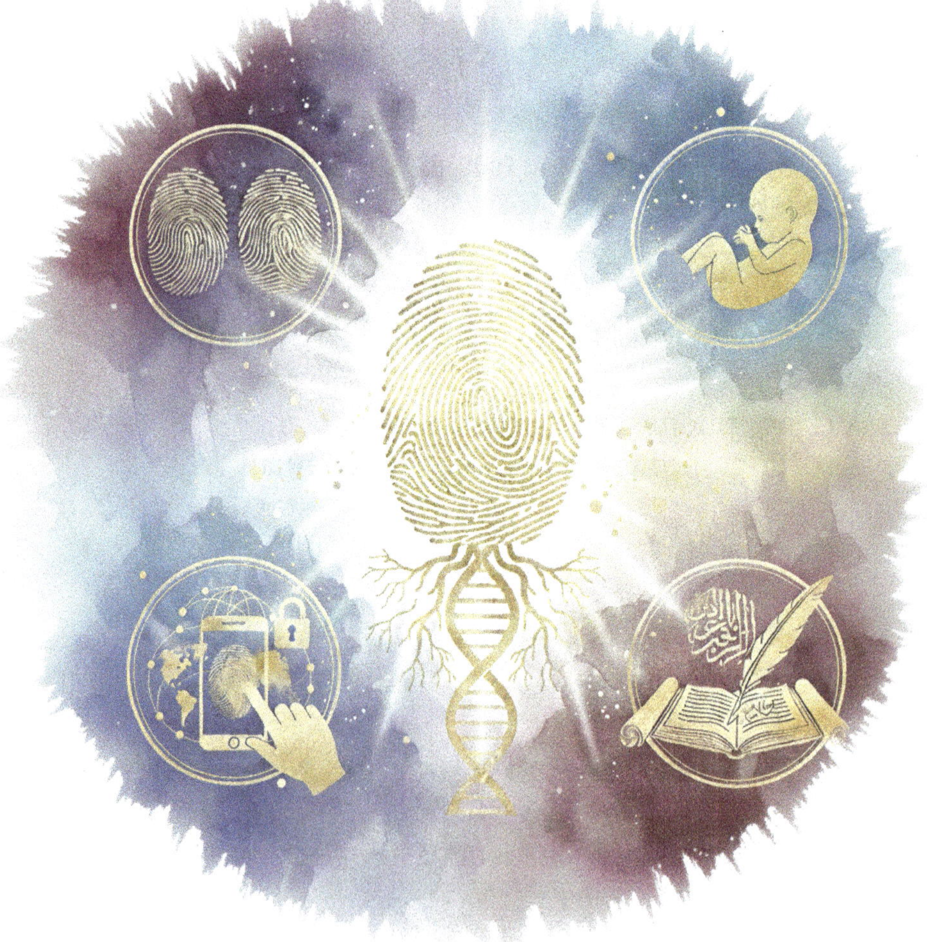

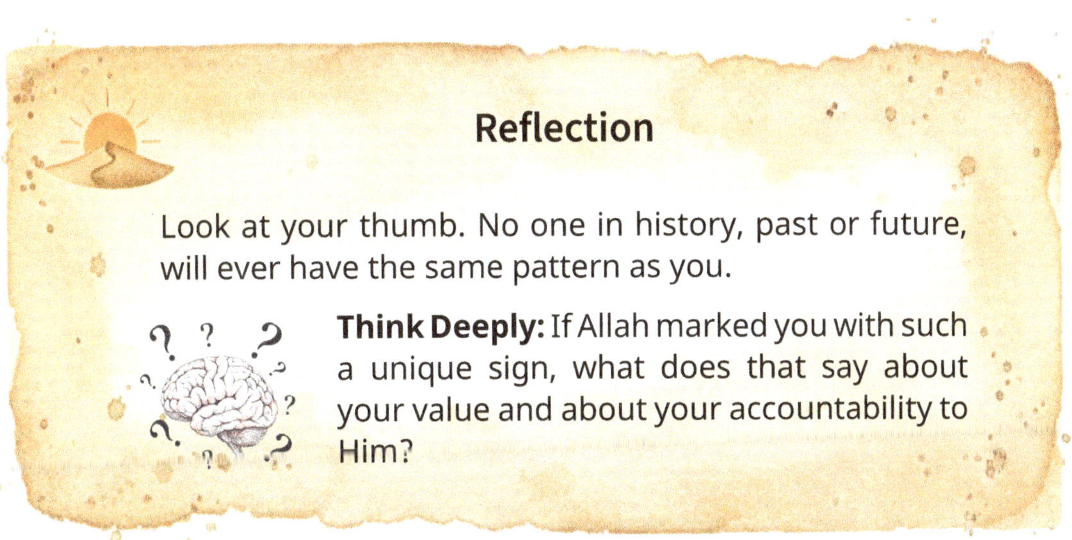

Reflection

Look at your thumb. No one in history, past or future, will ever have the same pattern as you.

Think Deeply: If Allah marked you with such a unique sign, what does that say about your value and about your accountability to Him?

The thumbprint is a small detail, but it carries a big message. It is a reminder that Allah creates with precision, knows us individually, and will resurrect us exactly as we were.

> **O Allah, just as You have perfected my creation, perfect my character.**

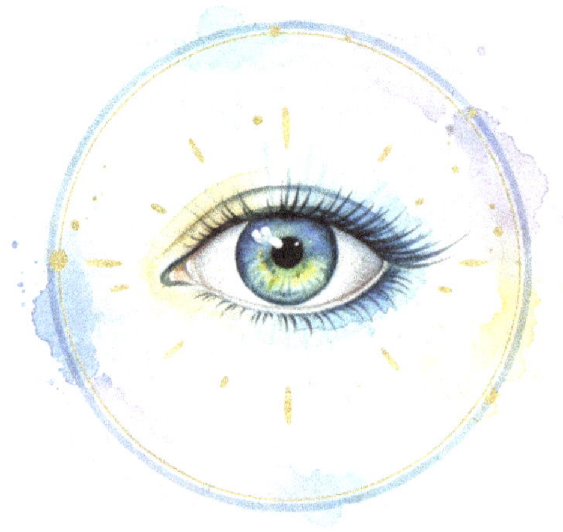

CHAPTER 9

THE HUMAN EYE
A WINDOW TO THE WORLD

It means: *"Say: It is He who created you and gave you hearing, sight, and hearts; little are you grateful."*
(Al-Mulk, 67:23)

The Gift of Sight

Every day, without thinking, we open our eyes and see the world. The Qur'an reminds us that it is Allah who gave us hearing, sight, and hearts. The eye is not only an organ; it is a wondrous sign. It lets us see colors, shapes, and movement. It helps us read, learn, and recognize the faces of those we love.

The Science of the Eye

The human eye is more advanced than any camera:

Light and Focus:

The cornea and lens bend light to focus images on the retina at the back of the eye.

Photoreceptors:

The retina contains about 120 million rods for light and dark vision and 6 million cones for color vision.

Instant Processing:

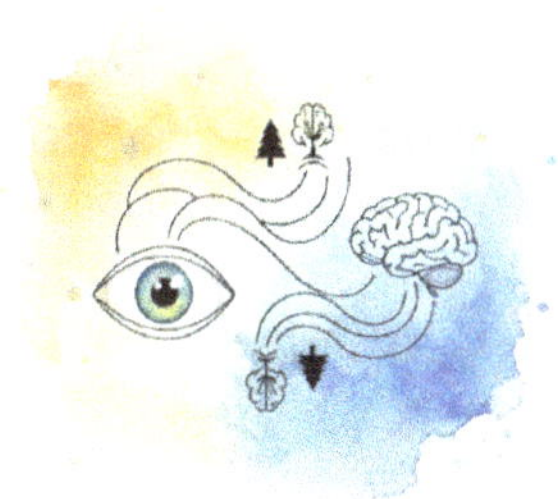

Signals travel through the optic nerve to the brain, which flips the upside-down image right side up.

Complex Coordination:

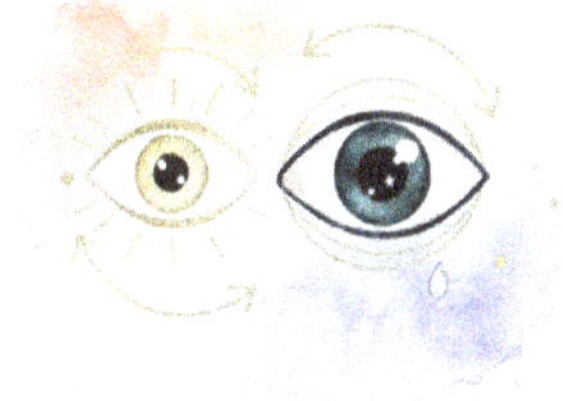

Muscles allow us to move our eyes smoothly in all directions. Blinking spreads tears to protect and clean them.

Adaptability:

Eyes adjust instantly between bright sunlight and dim darkness.

Despite modern technology, no camera comes close to the complexity and flexibility of the human eye.

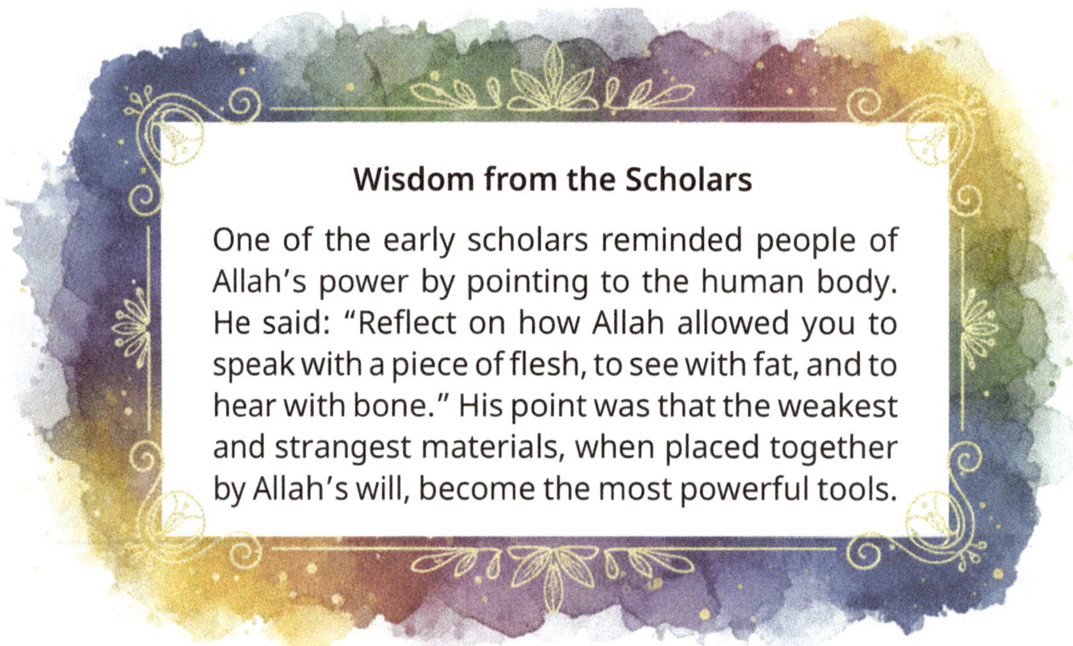

Wisdom from the Scholars

One of the early scholars reminded people of Allah's power by pointing to the human body. He said: "Reflect on how Allah allowed you to speak with a piece of flesh, to see with fat, and to hear with bone." His point was that the weakest and strangest materials, when placed together by Allah's will, become the most powerful tools.

Lessons from the Eye

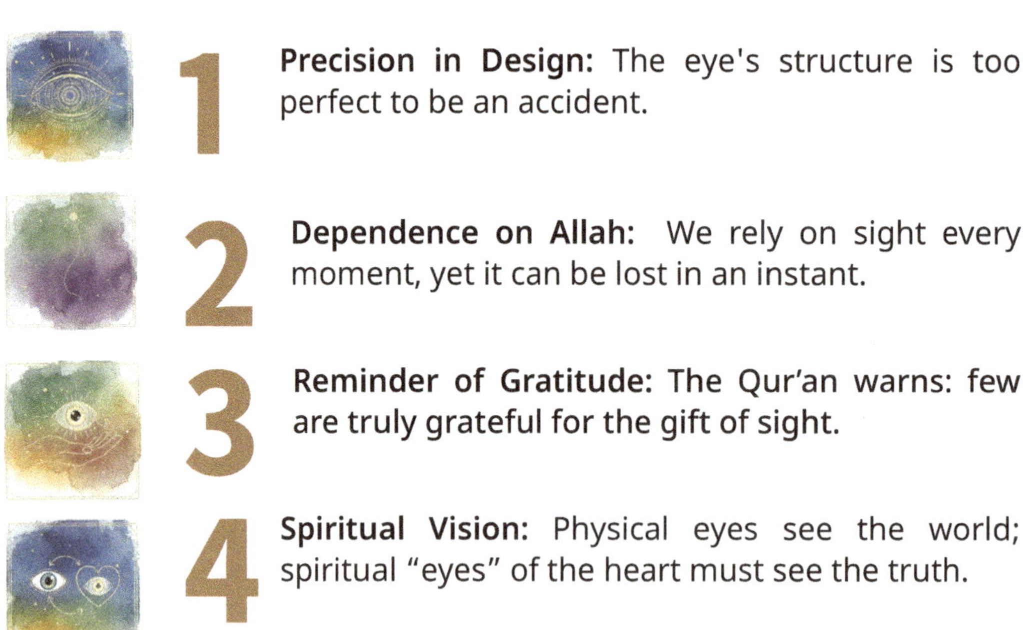

1. **Precision in Design:** The eye's structure is too perfect to be an accident.

2. **Dependence on Allah:** We rely on sight every moment, yet it can be lost in an instant.

3. **Reminder of Gratitude:** The Qur'an warns: few are truly grateful for the gift of sight.

4. **Spiritual Vision:** Physical eyes see the world; spiritual "eyes" of the heart must see the truth.

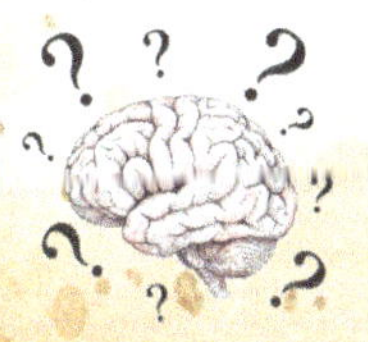

Reflection

Think about how much of your daily life depends on sight—reading messages, watching videos, navigating the world. Without your eyes, life would feel completely different.

Think Deeply: If Allah gave you physical eyes to see creation, shouldn't you also use your heart to know the Creator?

The human eye is tiny, delicate, yet powerful enough to take in the universe. It is one of Allah's greatest gifts, a sign that He created us with purpose and precision.

❝ O Allah, keep our eyes healthy, and open the eyes of our hearts to see the truth. ❞

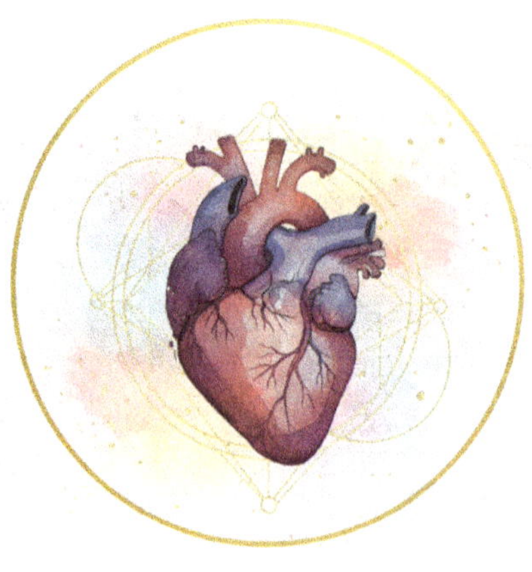

CHAPTER 10

THE HEART
BEATING WITHOUT REST

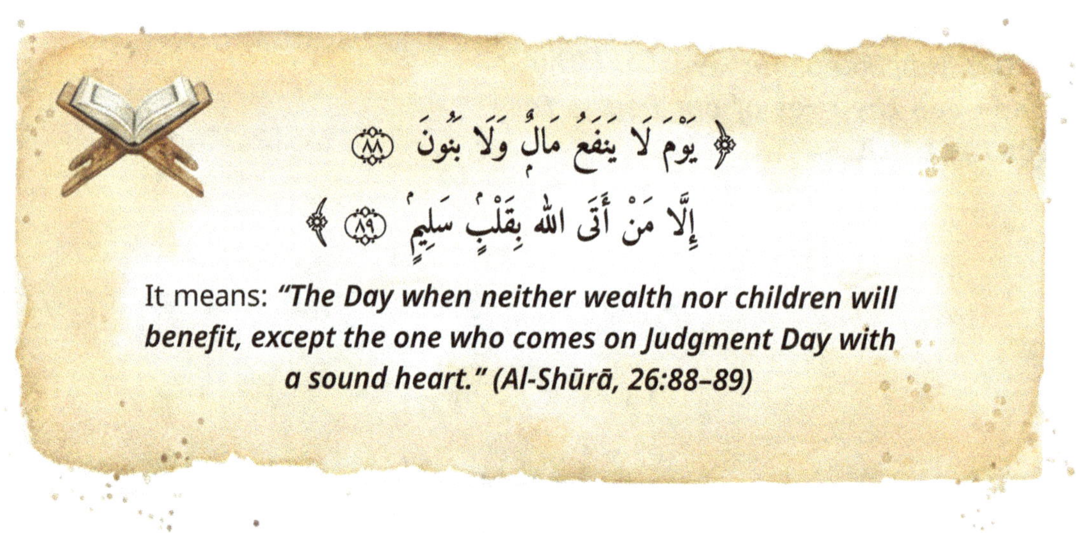

﴿ يَوْمَ لَا يَنْفَعُ مَالٌ وَلَا بَنُونَ ۝

إِلَّا مَنْ أَتَى اللهَ بِقَلْبٍ سَلِيمٍ ۝ ﴾

It means: *"The Day when neither wealth nor children will benefit, except the one who comes on Judgment Day with a sound heart." (Al-Shūrā, 26:88–89)*

The Most Vital Organ

No larger than a clenched fist, the heart beats continuously, day and night, without rest. Small in size yet immense in importance, it sustains every cell in the body. Just as the health of the physical heart is essential for life, the health of the heart within—the center of our intentions and faith—shapes the course of our lives.

The Science of the Heart

The human heart is an astonishing organ:

Constant Work:

It beats about 100,000 times a day

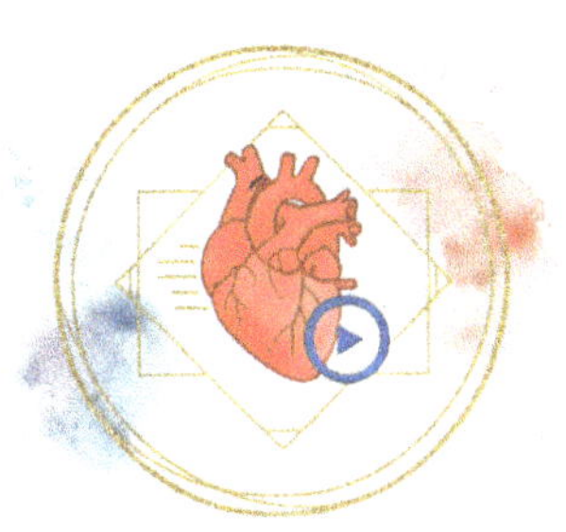

Blood Circulation:

It pumps about 5 liters of blood every minute. The blood travels through a network of roughly 100,000 km of blood vessels—a distance that could circle the earth twice.

Strength:

In an average lifetime, the heart will beat more than 3 billion times.

Dependence:

If the heart stopped for just a few minutes, life would end.

The heart's design is simple in shape but unmatched in endurance. No machine humanity has ever made runs so powerfully, so efficiently, and for so long without rest.

The Heart of Life

Long before science studied blood circulation, people knew the heart was special. It beats before we are born and continues without rest until we die. The Qur'an speaks often of the heart (qalb) with its deeper role, the spiritual aspect of it that understands, believes, and remembers Allah. The Qur'an teaches us that the heart is not only a muscle that pumps blood; it is also the center of understanding, awareness, and guidance.

Allah says:

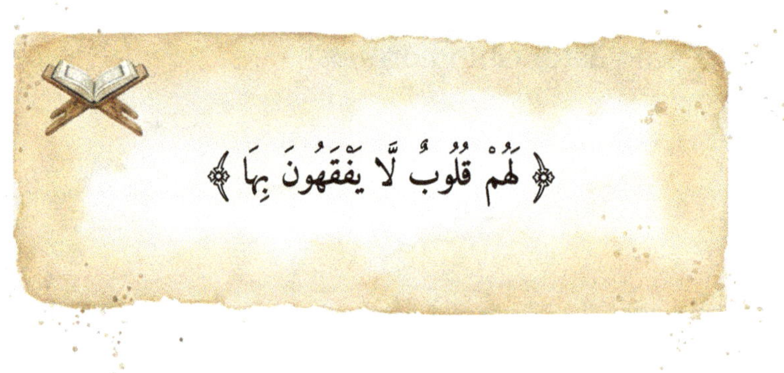

It means: ***"They have hearts with which they do not understand."***
(Al-Aʻrāf, 7:179)

And He says:

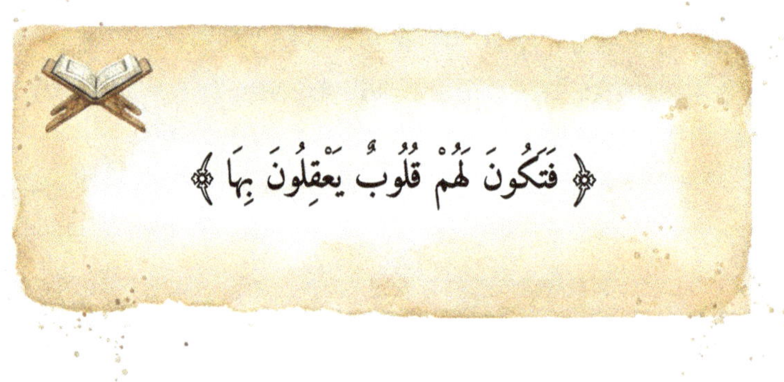

It means: ***"...so that they may have hearts with which they reason."***
(Al-Ḥajj, 22:46)

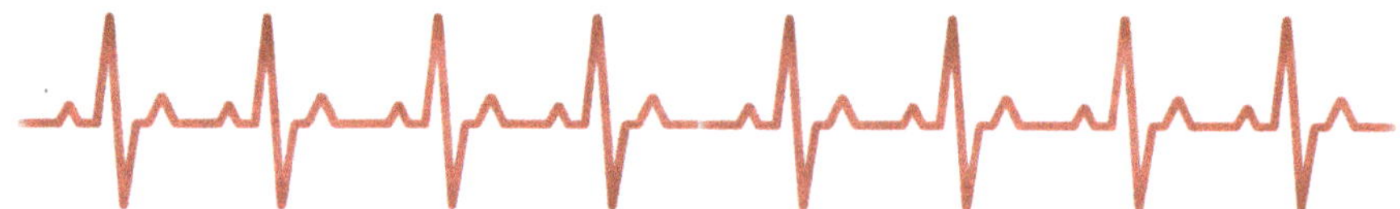

These verses show us that the heart is connected to our ability to recognize truth, reflect, and make sound judgments. In Islamic teaching, the physical heart and the spiritual heart are connected. The physical heart keeps us alive, while the spiritual heart shapes how we understand the world and respond to guidance. A clear and humble heart allows a person to benefit from knowledge, while a hardened heart can reject even the brightest truth.

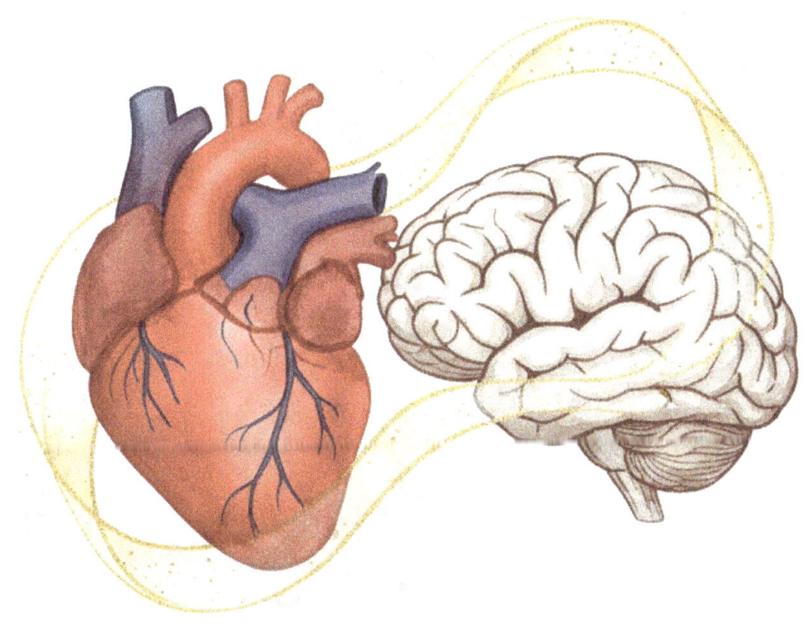

Prophetic wisdom

The heart is a small piece of flesh, yet it shapes everything we do. The Prophet ﷺ taught us that *"There is a piece of flesh in the body, if it is sound, the whole body is sound; and if it is corrupt, the whole body is corrupt. Truly, it is the heart."* Scholars explain that the heart is like a leader and the body follows it the way an army follows its commander. When the heart is sincere and upright, our actions shine with goodness. But when the heart becomes clouded by bad intentions or harmful desires, the rest of the body follows that direction as well.

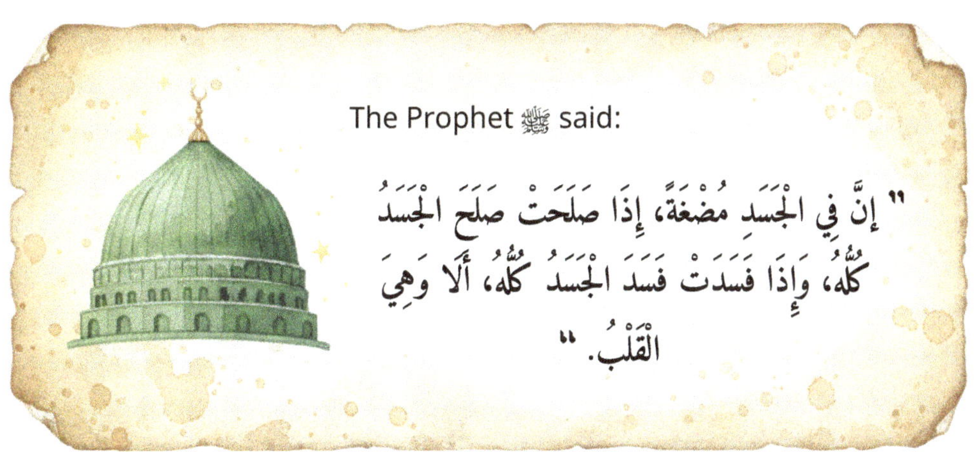

The Prophet ﷺ said:

Lessons from the Heart

1 **A Beating Reminder:** Every heartbeat is a sign of Allah's mercy keeping us alive.

2 **Fragility of Life:** Our lives depend on a small organ that can stop at any time.

3 **Two Kinds of Health:** Medical health keeps the heart pumping; spiritual health keeps it pure before Allah.

4 **Accountability:** On the Day of Judgment, wealth and status won't matter. Only a sound heart will.

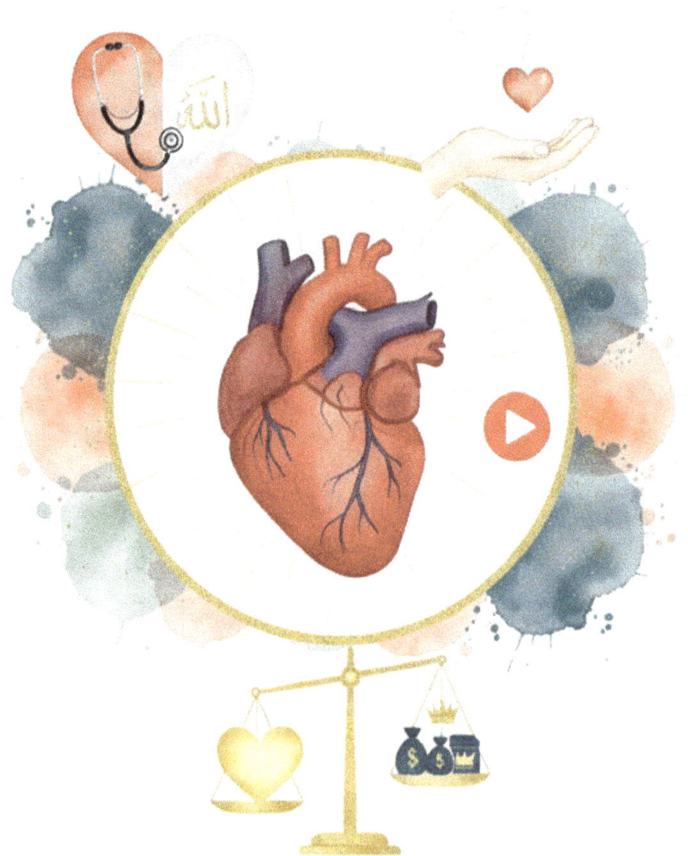

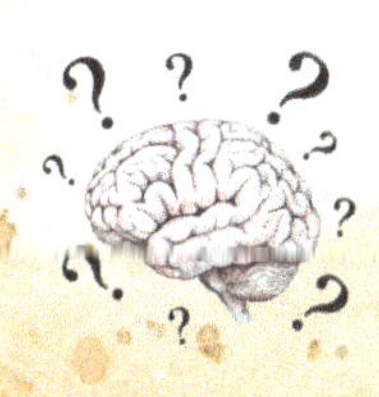

Reflection

Your heart never stops working, not even while you sleep. Every beat is a gift from Allah.

Think Deeply: If your physical heart needs care through healthy living, what about your spiritual heart? How do you feed it, cleanse it, and keep it alive with remembrance ot Allah?

The heart is both a pump and a symbol. It is a symbol for life in this world and represents our state in the next. Caring for both aspects of the heart, the physical and the spiritual, is part of faith.

❝ *O Allah, purify our hearts and let every beat remind us of You.* ❞

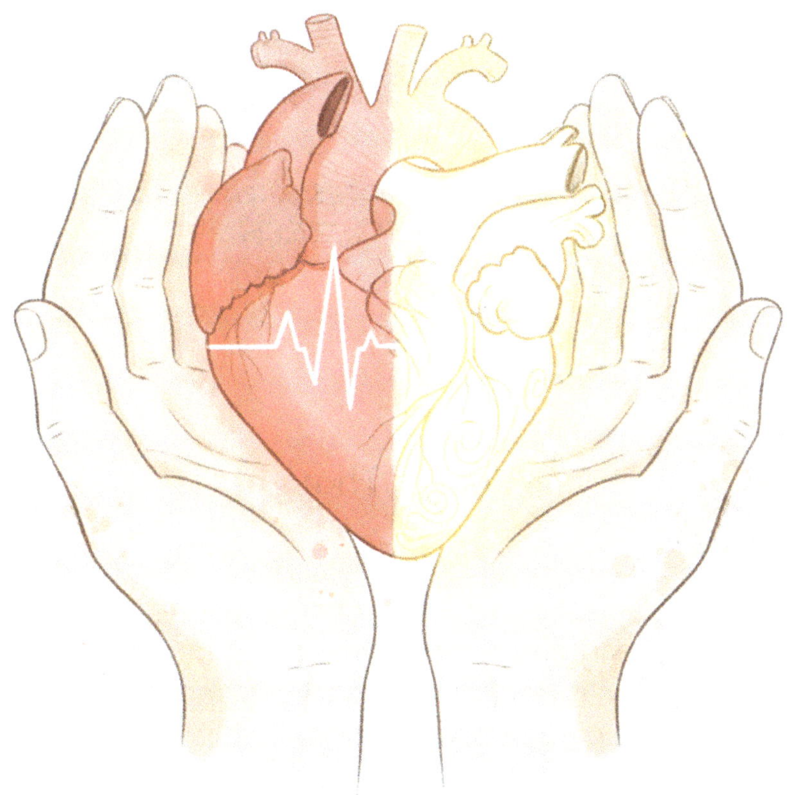

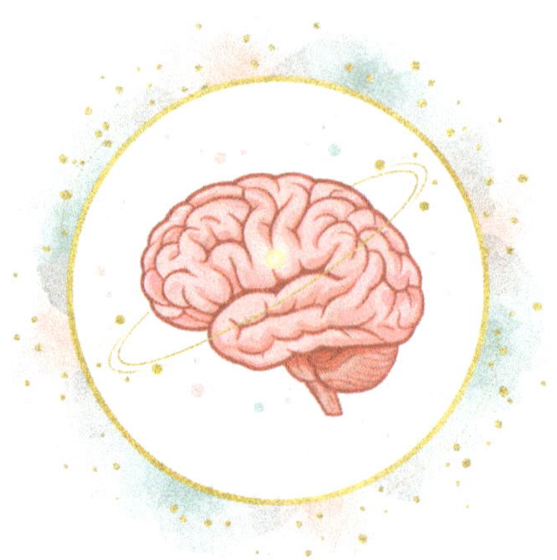

CHAPTER 11

THE BRAIN
MORE COMPLEX THAN A COMPUTER

The Seat of Thought

The Qur'an repeatedly asks us to use our minds—to think, reflect, and reason. The human heart is the place where that happens in conjunction with the brain. The brain is not just an organ; it is part of the control center of our body, the source of memory, creativity, and problem-solving.

The Science of the Brain

Neurons:

The brain contains about 86 billion nerve cells (neurons), many of them connected to thousands of others, forming trillions of connections.

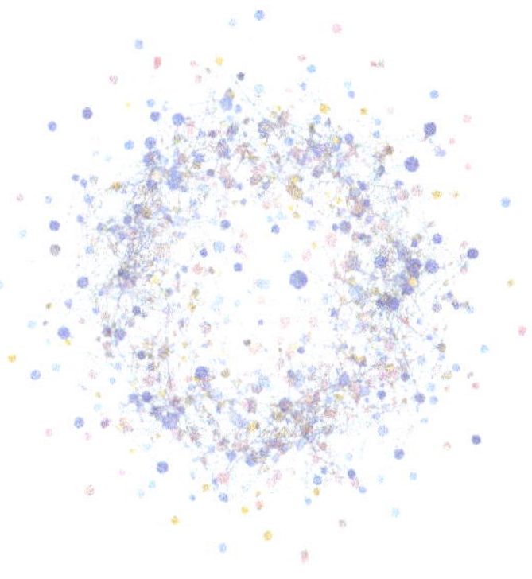

Speed

Signals travel at up to 400 km/h.

Memory:

The brain can store vast amount of information, far more than we can fully measure.

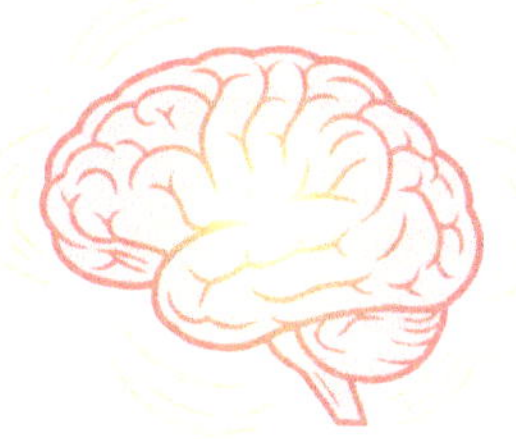

Energy:

Though only about 2% of body weight, the brain uses 20% of the body's energy.

Adaptability:

The brain can rewire itself, recovering after injury, or strengthening pathways through learning.

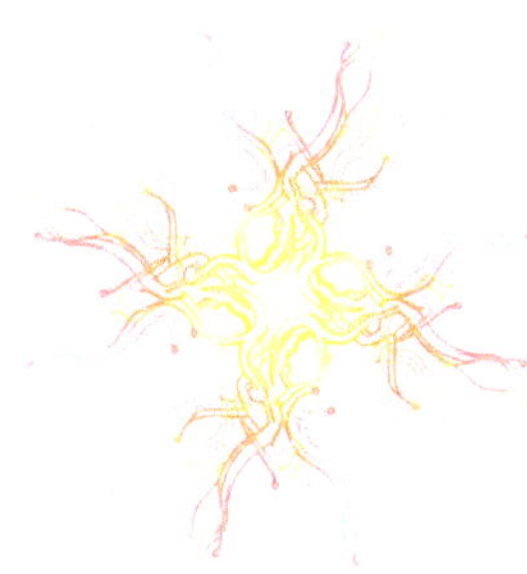

No machine humans have ever made comes close to the complexity of the brain.

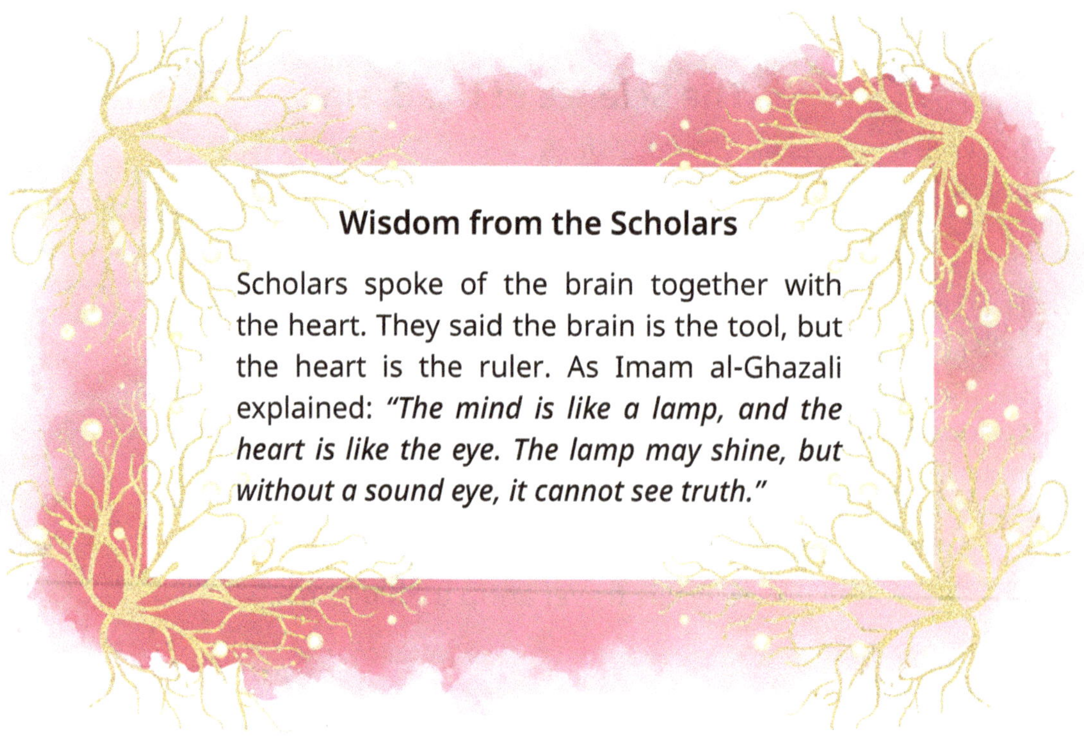

Wisdom from the Scholars

Scholars spoke of the brain together with the heart. They said the brain is the tool, but the heart is the ruler. As Imam al-Ghazali explained: *"The mind is like a lamp, and the heart is like the eye. The lamp may shine, but without a sound eye, it cannot see truth."*

Lessons from the Brain

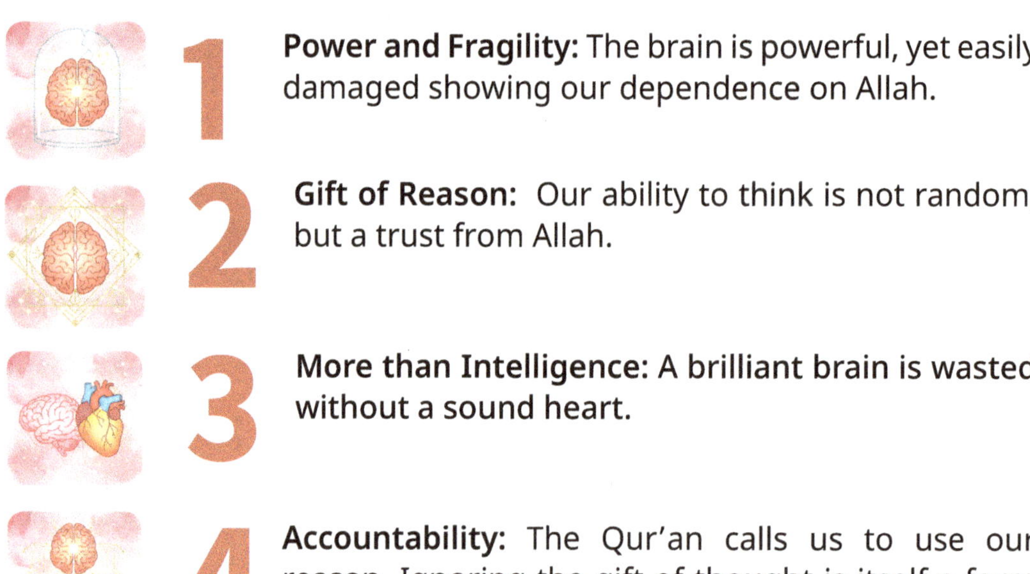

1 **Power and Fragility:** The brain is powerful, yet easily damaged showing our dependence on Allah.

2 **Gift of Reason:** Our ability to think is not random, but a trust from Allah.

3 **More than Intelligence:** A brilliant brain is wasted without a sound heart.

4 **Accountability:** The Qur'an calls us to use our reason. Ignoring the gift of thought is itself a form of ingratitude.

Reflection

Your brain processes thoughts with astonishing speed and complexity, yet it cannot even keep itself alive for a single moment without Allah's will.

Think Deeply: Do you use your brain only for school and work or also to reflect on Allah's signs and your purpose?

The brain is full of power, memory, and creativity. But it is also fragile, reminding us that intelligence alone is not enough. True guidance is when the brain works with a heart alive in faith.

> **O Allah, make our minds sharp in understanding, and our hearts firm in belief.**

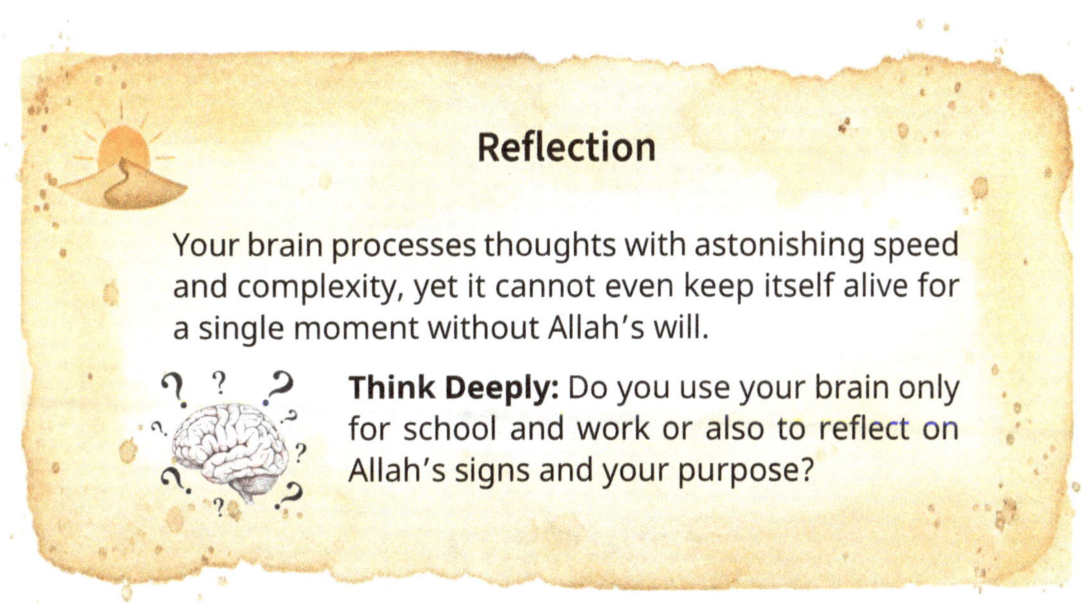

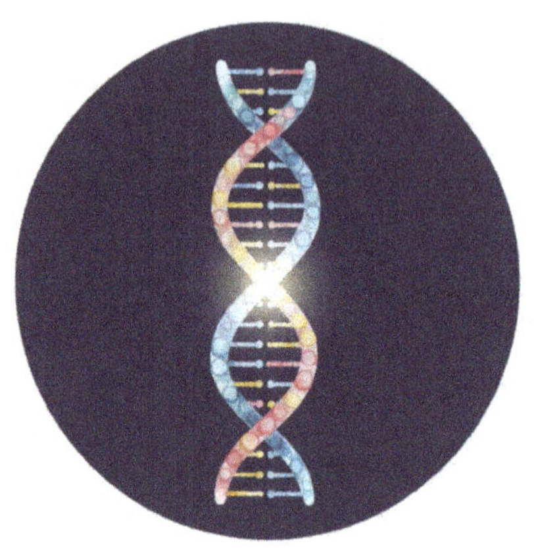

CHAPTER 12

THE DNA
A CODE WITHIN EVERY CELL

It means: *"Indeed, We created man from a drop of mingled fluid to test him; and We gave him hearing and sight."*
(Al-Insān, 76:2)

The Hidden Code Inside Us

Inside every cell of your body is a complete set of instructions represented by scientists in a language of four "letters": A, T, C, and G. This is DNA, the code that tells your cells how to grow, function, and repair themselves. It is like a library written inside you, present in every single cell.

The Science of the DNA

Length:

Scientists say if all the DNA in your body were uncoiled, it would stretch from the earth to the sun and back many times.

Precision

DNA contains more than three billion chemical bases arranged in a precise sequence. This sequence is part of what makes you who you are.

Uniqueness:

No two people share the same DNA sequence; even identical twins have small genetic differences.

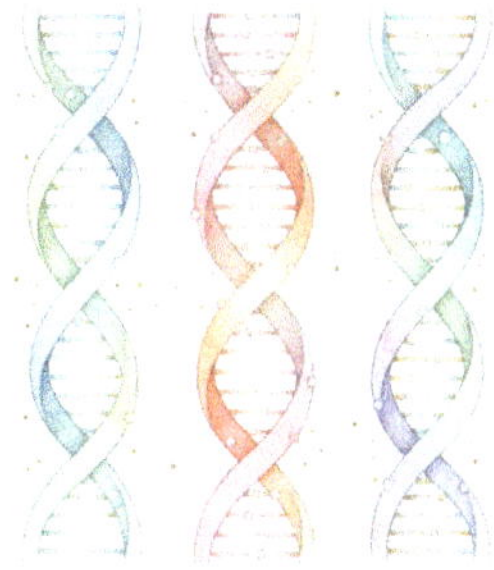

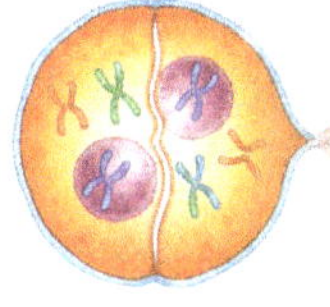

Replication:

Every time your cells divide, the DNA copies itself with incredible accuracy. Errors are rare, though even they are part of Allah's decree.

DNA is like a signature—an unseen code contributing to what's makes you unique.

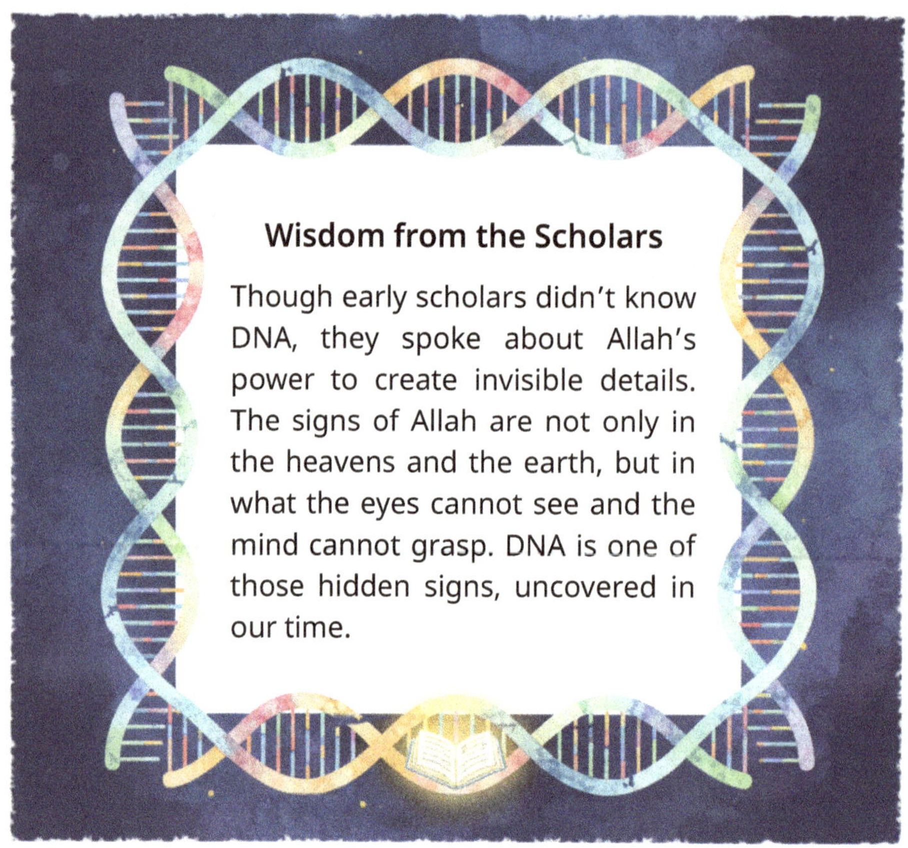

Lessons from the DNA

1. Allah's Knowledge Is Complete: your genetic code is written before you are born..

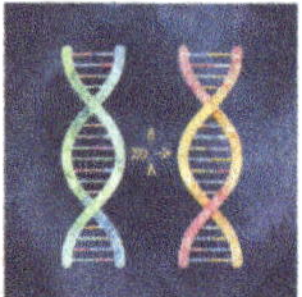

No Two are Alike: Your DNA proves Allah creates each person individually.

Order and Precision: DNA shows that life is not random but follows a remarkably precise order.

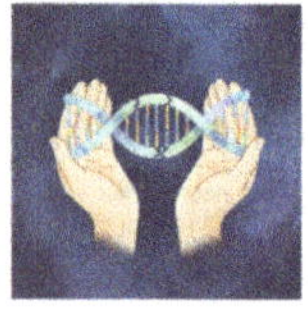

Dependence on Allah: Even the tiniest errors in DNA can cause illness, reminding us of our fragility.

Reflection

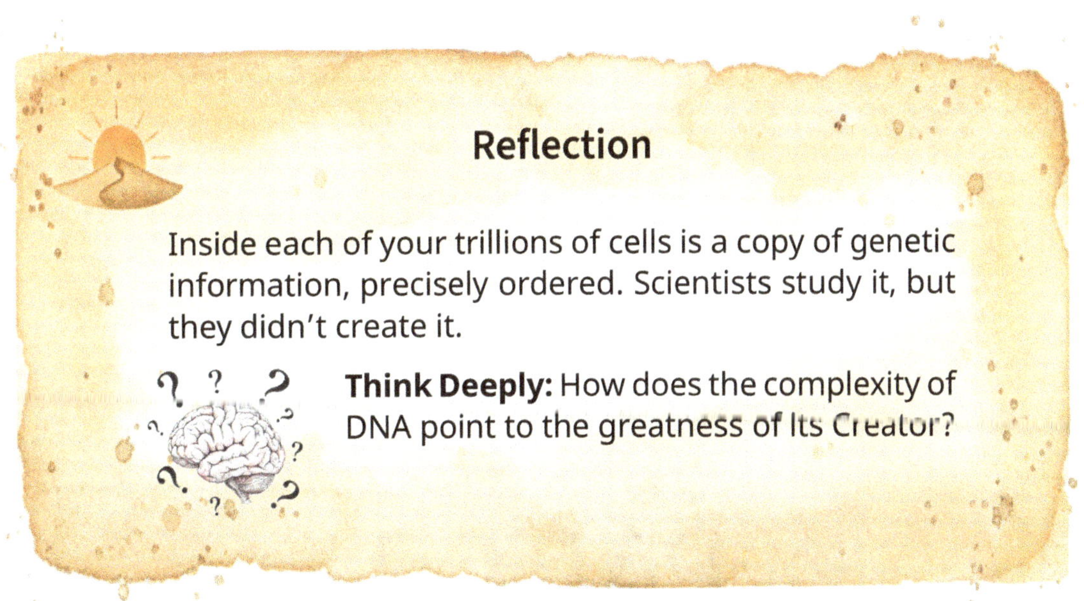

Inside each of your trillions of cells is a copy of genetic information, precisely ordered. Scientists study it, but they didn't create it.

Think Deeply: How does the complexity of DNA point to the greatness of Its Creator?

DNA is invisible to the naked eye, yet it carries information essential for life. It is a reminder that Allah's wisdom reaches into the smallest details of creation.

> *O Allah, You created us with precision and knowledge. Let us live in a way that honors the purpose we are created for.*

PART 3

SIGNS IN THE HEAVENS & EARTH

The existence of the universe, the earth, the sky, and human beings point to a Creator who created all of these with wisdom and power. Through reflection, a person comes to know he has a Creator who brought him into existence and created the heavens and the earths. Creation does not create itself. Plants did not create themselves, nor did the stars or crops. This is a rational proof of the existence of Allah.

CHAPTER 13

THE SKY
A CANOPY ABOVE US

It means: *"And We made the sky a protected ceiling, but they turn away from its signs."* (Al-Anbiyāʾ, 21:32)

The Sky Above Us

Every time we look up, we see the sky stretching above far than the eyes can see. For most people, it feels ordinary, but the Qur'an tells us it is one of the greatest signs of Allah. The sky is not just beautiful; it is a shield and a reminder of the Creator's mercy. The sky contains wonders that indicate Allah's power, wisdom, and knowledge.

The Science of the Sky

Protection:

The atmosphere acts as a shield, burning up meteors before they strike earth. Without it, the planet would be constantly bombarded.

Oxygen and Life:

The air around us carries the oxygen we breathe and the carbon dioxide plants need.

Ozone Layer:

It filters the sun's harmful rays, while allowing light and warmth that life depends on.

Weather and Water:

Clouds form in the sky, carrying water that falls as rain, bringing life to the earth.

Vastness:

Scientists still cannot measure the universe's full extent. The more we learn, the more we realize our smallness beneath this vast canopy.

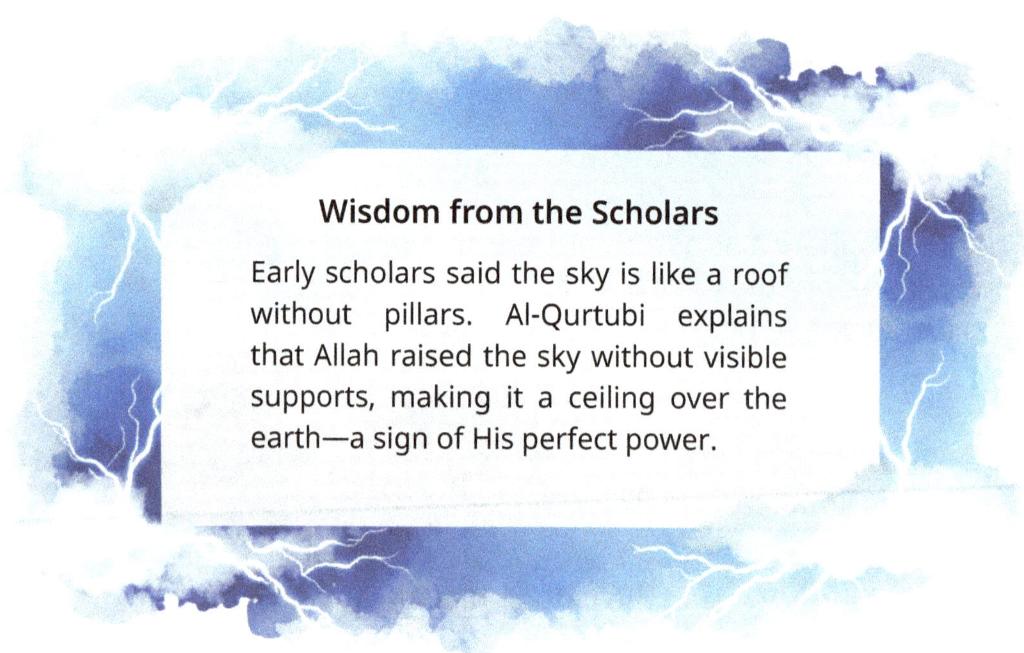

Lessons from the Sky

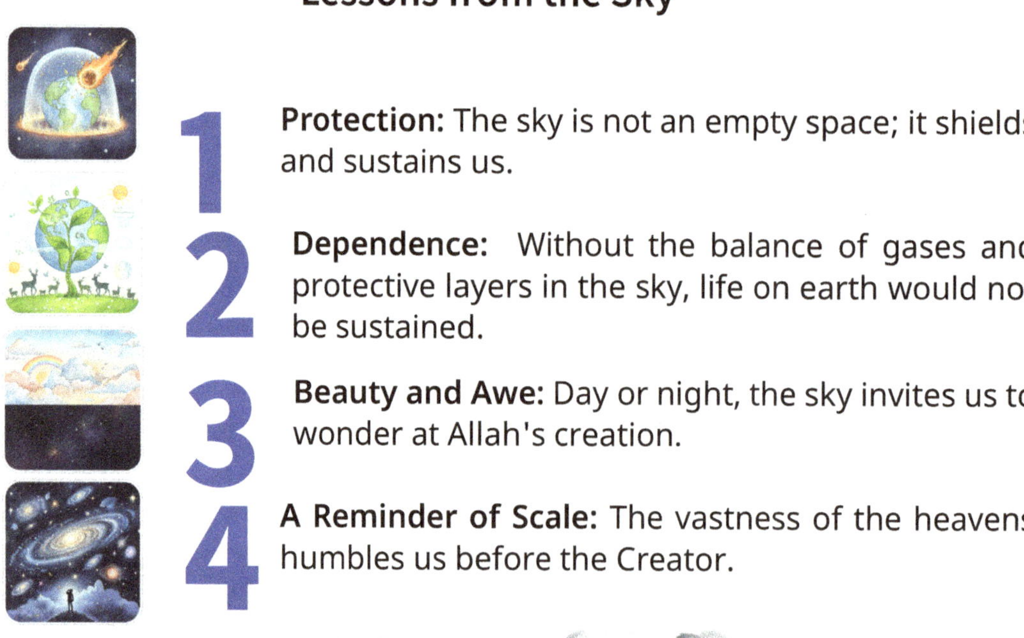

1 **Protection:** The sky is not an empty space; it shields and sustains us.

2 **Dependence:** Without the balance of gases and protective layers in the sky, life on earth would not be sustained.

3 **Beauty and Awe:** Day or night, the sky invites us to wonder at Allah's creation.

4 **A Reminder of Scale:** The vastness of the heavens humbles us before the Creator.

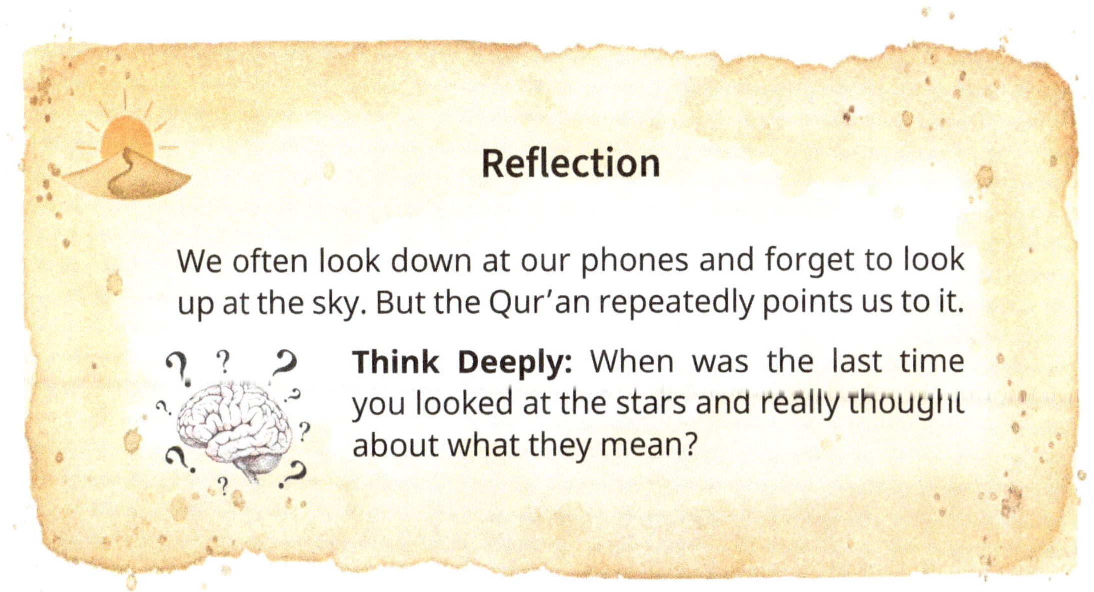

Reflection

We often look down at our phones and forget to look up at the sky. But the Qur'an repeatedly points us to it.

Think Deeply: When was the last time you looked at the stars and really thought about what they mean?

The sky is more than scenery. It is a ceiling, a shield, and a sign. Every breath we take and every glimpse of the stars reminds us of Allah's care and power.

" *O Allah, let the sky remind us of Your greatness, and keep our hearts from turning away from Your signs.* **"**

CHAPTER 14

THE MOUNTAINS
PEGS OF THE EARTH

﴿ أَلَمْ نَجْعَلِ ٱلْأَرْضَ مِهَٰدًا ۝ وَٱلْجِبَالَ أَوْتَادًا ۝ ﴾

It means: *"Have We not made the earth a safe, comfortable and spread out place to live like a bed prepared for someone to rest on, and the mountains as pegs?"* (Al-Nabaʾ, 78:6–7)

The Strength of Mountains

Mountains rise high above the earth, yet the Qur'an describes them as awtād—pegs or stakes. Allah has anchored them on the earth's surface as stabilizers. Just as a tent peg holds a tent firm in place, mountains are signs of stability for the earth. Their beauty and majesty make us stop and reflect. They are not random rock formations, but part of Allah's perfect order for the Earth.

The Science of Mountains

Stability:

Mountains have deep roots beneath the earth, anchoring the crust like pegs.

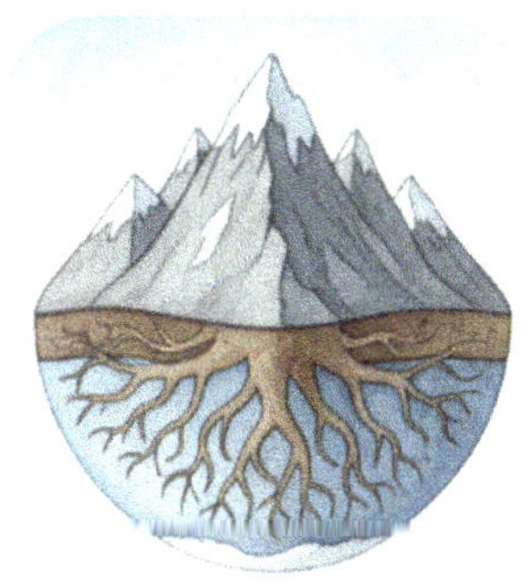

Water Sources:

Snow and ice on mountains melt to provide rivers and fresh water for millions of people.

Habitats:

Mountains are home to unique plants and animals adapted to high altitudes.

Climate Effect:

They block winds, shape rainfall, and produce valleys of life.

Age and Endurance:

Some mountains are nearly as old as the Earth, standing firm through countless generations.

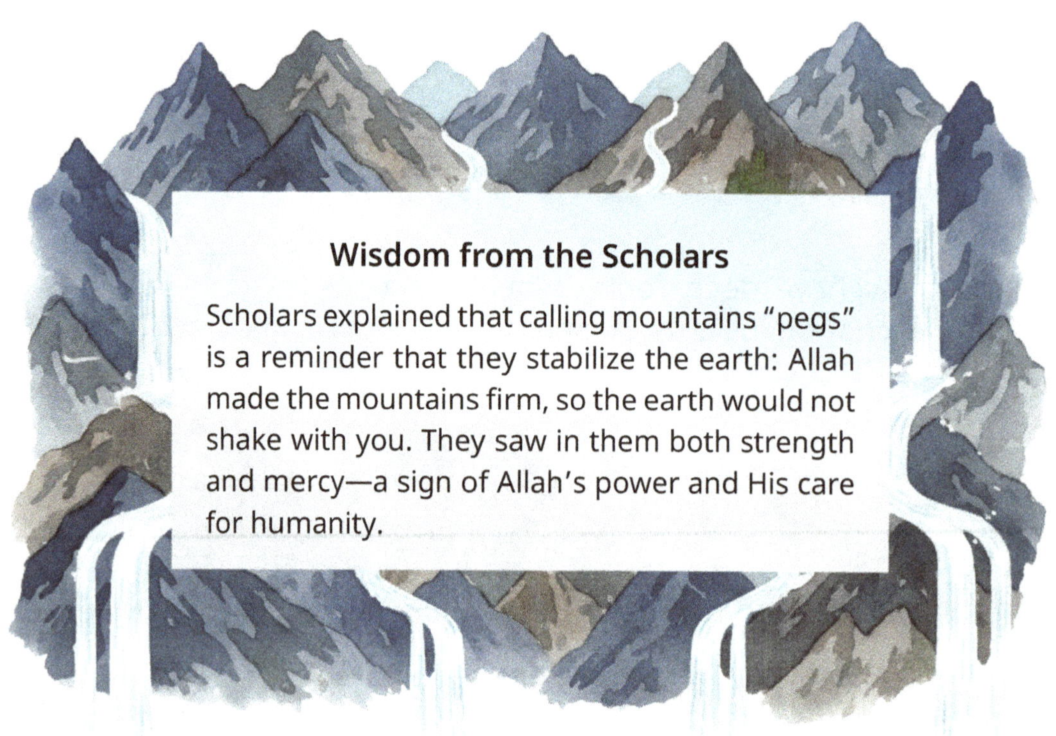

Wisdom from the Scholars

Scholars explained that calling mountains "pegs" is a reminder that they stabilize the earth: Allah made the mountains firm, so the earth would not shake with you. They saw in them both strength and mercy—a sign of Allah's power and His care for humanity.

Lessons from the Mountains

1 **Firmness and Stability:** Mountains remind us that Allah made the earth livable, not chaotic.

2 **Provision:** From them flow rivers, minerals, and fertile soil.

3 **Majesty:** Their size and beauty point to Allah's greatness.

4 **Reminder of Endurance:** Just as mountains outlast generations, our deeds endure beyond our short lives.

Reflection

Standing at the foot of a mountain, we feel small. That feeling should not make us afraid, but humble before the One who created both us and the mountains.

Think Deeply: If mountains stand tall for generations, how firm should your faith be in the short years of your life?

The mountains are both anchors of the earth and symbols of strength. They remind us that Allah made the world stable for us to live in and that our own lives should be built on the solid foundation of faith.

❝ *O Allah, make our hearts firm as You made the mountains firm upon the earth.* **❞**

CHAPTER 15

THE SUN AND THE MOON

MEASURES OF TIME

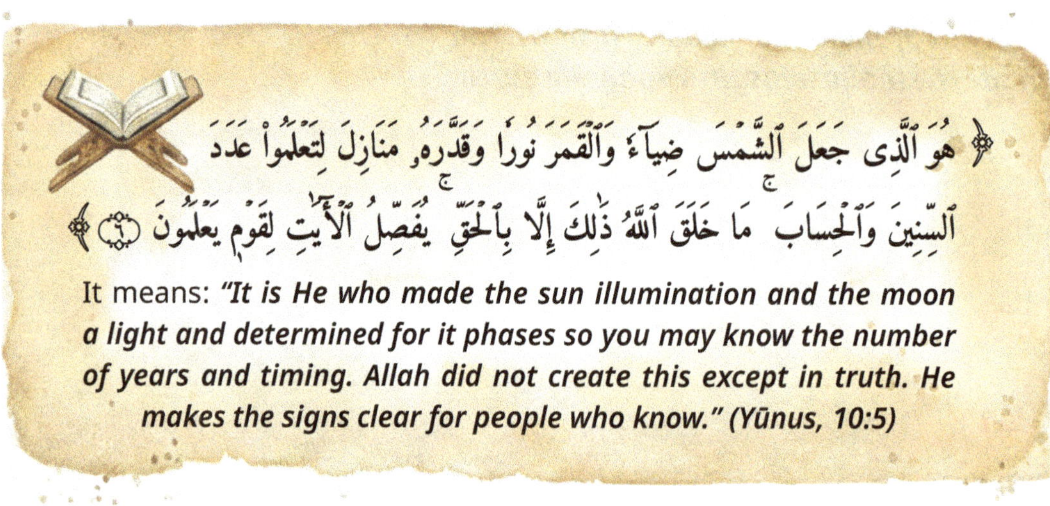

هُوَ ٱلَّذِى جَعَلَ ٱلشَّمْسَ ضِيَآءً وَٱلْقَمَرَ نُورًا وَقَدَّرَهُۥ مَنَازِلَ لِتَعْلَمُواْ عَدَدَ ٱلسِّنِينَ وَٱلْحِسَابَ ۚ مَا خَلَقَ ٱللَّهُ ذَٰلِكَ إِلَّا بِٱلْحَقِّ ۚ يُفَصِّلُ ٱلْآيَٰتِ لِقَوْمٍ يَعْلَمُونَ ۝

It means: *"It is He who made the sun illumination and the moon a light and determined for it phases so you may know the number of years and timing. Allah did not create this except in truth. He makes the signs clear for people who know." (Yūnus, 10:5)*

The Sun and the Moon in Our Lives

The sun and moon are constant companions of life on earth. The sun brings light, warmth, and energy by day. The moon shines by night, emitting light in gentle phases. Beyond their beauty, they are measures of time—days, months, and years—guiding people long before clocks were invented.

The Science of the Sun and the Moon

The Sun:

A visible creation that gives light and heat whose benefits extend to various form of life. It supports human health by helping our bodies produce vitamin D, which plays an important role in bone strength and overall wellness. It nourishes plants and trees, warms and regulates the atmosphere, purifies water and air, and spreads daylight that allows human life and daily affairs to function.

The moon:

Its predictable orbit produces phases that mark time and form the basis of lunar months, still central to the Islamic calendar today.

Precision:

Their orbits are precise. The sun never collides with the moon, and each moves in its own course, exactly as the Qur'an describes (Yā Sīn, 36:40).

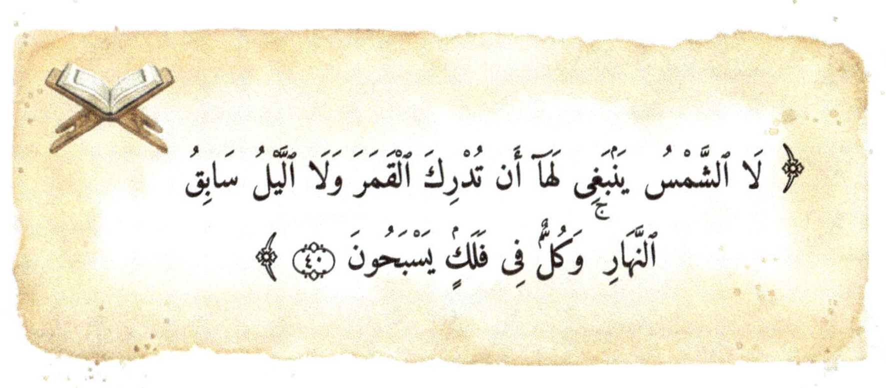

Lessons from the Sun and the Moon

1 **Balance and Order:** Their perfect cycles show Allah's control over time and creation.

2 **Guidance for Life:** Days, nights, months, and years are measured by them.

3 **Signs, Not Things to be Worshipped:** Some ancient people worshipped the sun and moon, but the Qur'an reminds us they are creations submitting to Allah.

4 **Dependence on Allah:** The sun and the moon remind us how dependent we are on the One who created them.

Reflection

Think of how much of your life is structured by the rising and setting of the sun, or by the months of the moon. These are not coincidences, but divine order.

Think Deeply: If the sun and moon obey Allah in perfect cycles, should we not also submit to Him with discipline and order?

The sun and the moon are signs of time, balance, and beauty. They remind us that Allah created everything with purpose and that our lives too must have purpose.

❝ *O Allah, make our days and nights filled with Your remembrance, as You made the sun and the moon signs of Your greatness.* ❞

CHAPTER 16

NIGHT AND DAY
ALTERNATING SIGNS

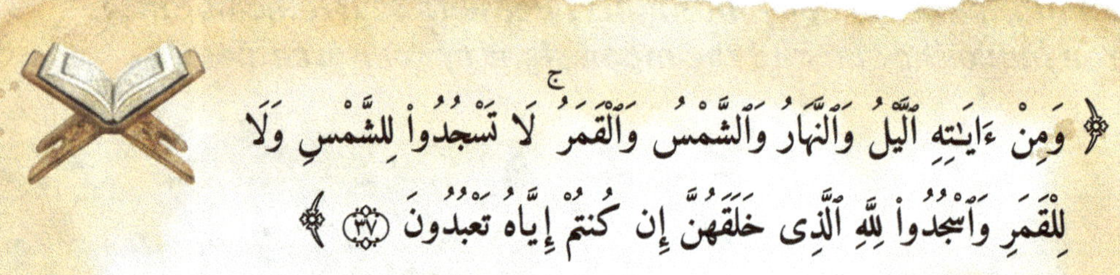

It means: *"And of His signs are the night and the day, and the sun and the moon. Do not prostrate to the sun or to the moon, but prostrate to Allah who created them, if it is truly Him you worship."*
(Fuṣṣilat, 41:37)

The Rhythm of Life

Night and day are so familiar that we hardly notice them. Yet they are among the greatest signs of Allah. Their perfect alternation shapes the rhythm of life on Earth. The Qur'an reminds us that they are not to be worshipped but creations—signs calling us to worship Allah alone.

The Science of Night and Day

Balance:

Allah created a perfect balance between day and night. If this balance were different, life on Earth would be difficult, Days could become unbearably hot while nights could turn freezing cold.

Rest and Work:

The darkness of night gives rest, and the brightness of day allows activity—a cycle perfectly suited for human life.

Photosynthesis:

Plants depend on this rhythm to produce oxygen, the air we breathe.

Time and Seasons:

The changing length of night and day through the year shapes our seasons, food cycles, and even our moods.

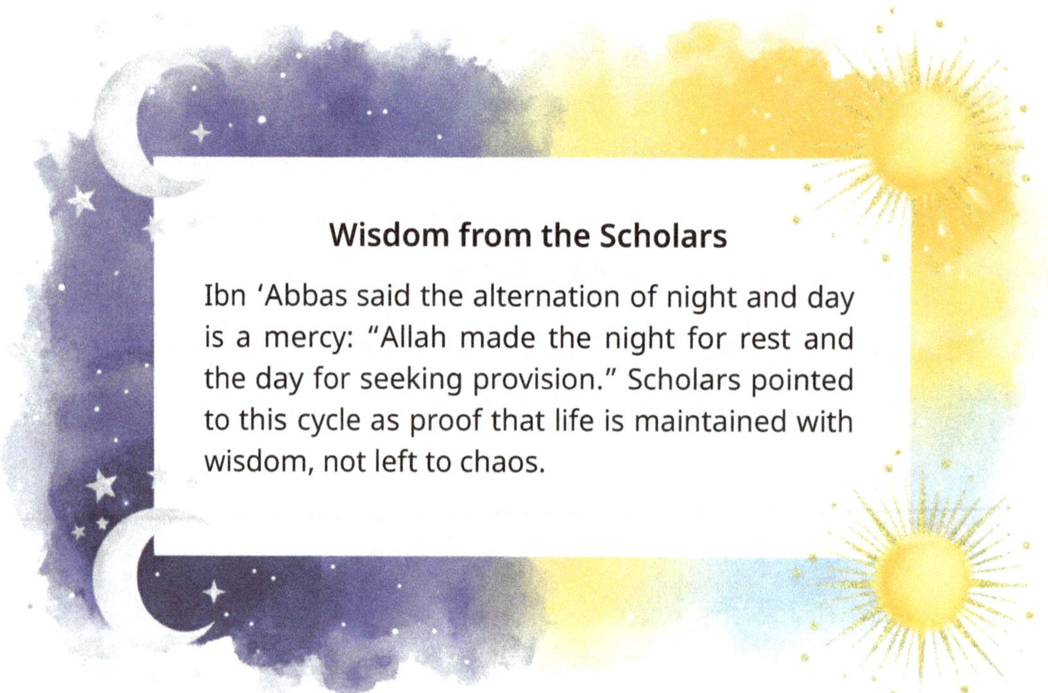

Wisdom from the Scholars

Ibn 'Abbas said the alternation of night and day is a mercy: "Allah made the night for rest and the day for seeking provision." Scholars pointed to this cycle as proof that life is maintained with wisdom, not left to chaos.

Lessons from Night and Day

1 **Mercy in Balance:** Night for sleep, day for work, each is a blessing.

2 **Reminder of Limits:** We cannot stop night from falling or day from rising; we are always under Allah's will.

3 **Worship of Allah Alone:** The Qur'an directs us to worship the One who made the night and the day succeed one another in perfect order, so that the night does not outrun the day.

4 **Symbol of Life:** Just as night is followed by day, difficulty is followed by ease.

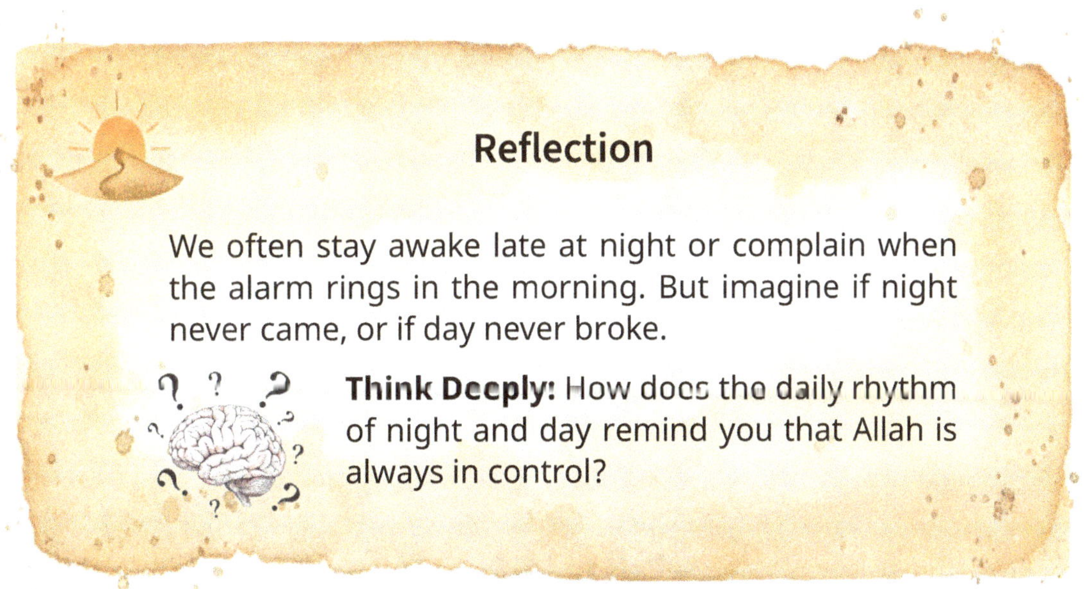

Reflection

We often stay awake late at night or complain when the alarm rings in the morning. But imagine if night never came, or if day never broke.

Think Deeply: How does the daily rhythm of night and day remind you that Allah is always in control?

The alternation of night and day is more than a routine—it is a sign, a mercy, and a reminder of the Creator's wisdom.

" O Allah, let our nights be filled with rest and worship, and our days with energy and gratitude. "

CHAPTER 17

RAIN AND PLANTS
LIFE FROM WATER

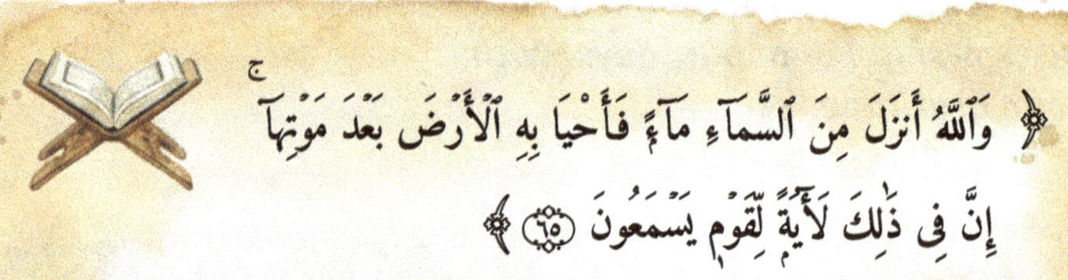

وَٱللَّهُ أَنزَلَ مِنَ ٱلسَّمَاءِ مَآءً فَأَحْيَا بِهِ ٱلْأَرْضَ بَعْدَ مَوْتِهَآ

إِنَّ فِي ذَٰلِكَ لَآيَةً لِّقَوْمٍ يَسْمَعُونَ ٦٥

It means: *"And Allah sends down water from the sky and thereby gives life to the earth after its death. Indeed, in that is a sign for people who listen."* (Al-Naḥl, 16:65)

Water from the Sky

Rain is a mercy from Allah. Dry land becomes green, seeds burst into life, and animals and people are nourished. Rain is not random; it is part of a system perfectly made to sustain life on earth. In the sky is our provision, meaning the rain which is a cause of our sustenance.

Science of Rain and Plants

Diversity from One Water:

From the same rain, Allah brings out countless varieties of fruits, grains, and plants, each with its own taste, color, and benefit.

Purification:

Rain cleans the air and refreshes the land.

Balance:

Too much water causes floods; too little causes drought—the balance is by Allah's will.

Dependence:

Many things, from the smallest insect to the largest tree, depends on water.

Wisdom from the Scholars

Imam al-Tabari explained that when Allah says He gives "life to the earth after its death," it means both the visible greening of plants and the deeper reminder of resurrection—just as the dead land comes to life, so will humans be brought back to life on the Day of Judgment.

Lessons from Rain and Plants

1 **Mercy and Provision:** Rain is a direct gift from Allah, bringing food and life.

2 **Variety from Unity:** From one kind of water, come countless different crops.

3 **Reminder of Resurrection:** The greening of the earth is a sign of life after death.

4 **Human Limits:** Despite all technology, only Allah controls the rain.

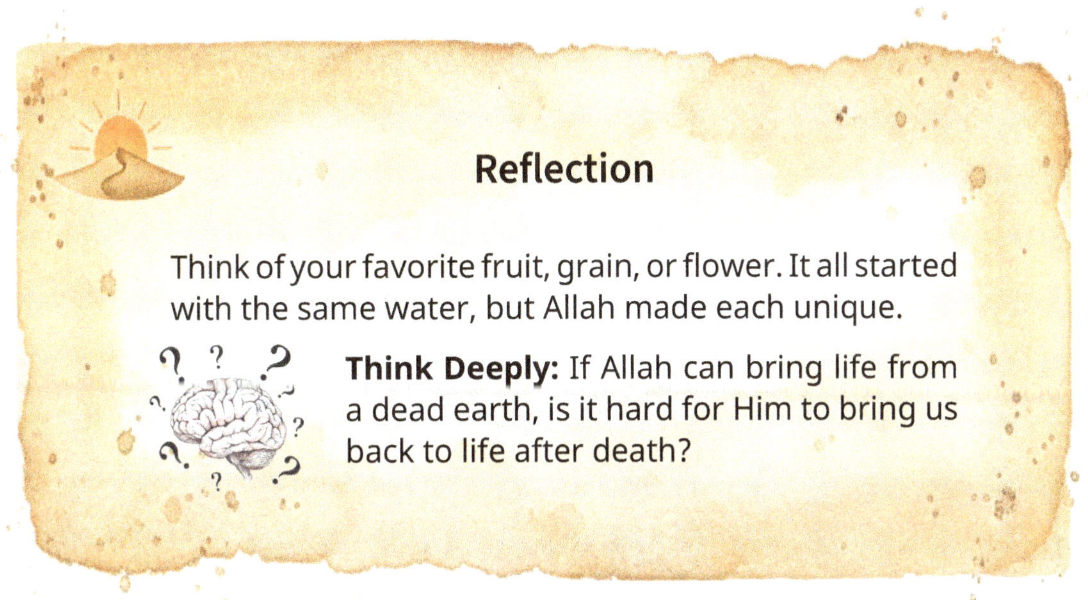

Reflection

Think of your favorite fruit, grain, or flower. It all started with the same water, but Allah made each unique.

Think Deeply: If Allah can bring life from a dead earth, is it hard for Him to bring us back to life after death?

Rain and plants are not just parts of nature; they are signs of mercy, balance, and resurrection. Every drop is a reminder of Allah's power and care.

> **O Allah, send us rain of mercy, not of punishment, and let us be among those who are revived with Your remembrance.**

CHAPTER 18

THE EARTH
SPREAD OUT FOR YOU

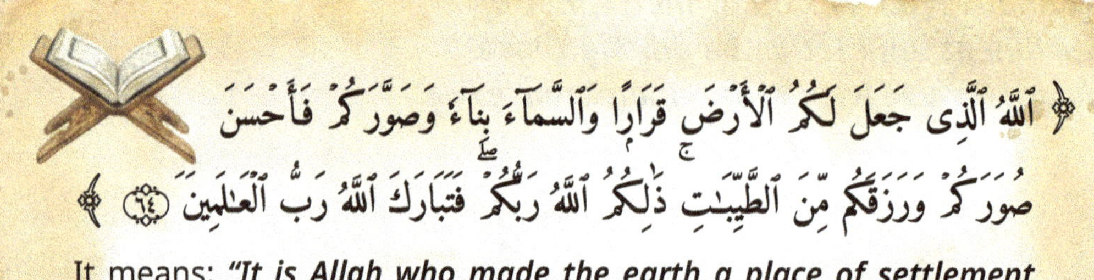

ٱللَّهُ ٱلَّذِى جَعَلَ لَكُمُ ٱلْأَرْضَ قَرَارًا وَٱلسَّمَاءَ بِنَاءً وَصَوَّرَكُمْ فَأَحْسَنَ صُوَرَكُمْ وَرَزَقَكُم مِّنَ ٱلطَّيِّبَٰتِ ذَٰلِكُمُ ٱللَّهُ رَبُّكُمْ فَتَبَارَكَ ٱللَّهُ رَبُّ ٱلْعَٰلَمِينَ ٦٤

It means: *"It is Allah who made the earth a place of settlement for you and the sky a canopy, and formed you, and perfected your forms and provided you with good things. That is Allah, your Lord; praised is Allah, Lord of the worlds." (Surah Ghāfir, 40:64)*

A Home for Humanity

The earth is not another planet among many. It is a prepared home. Allah spread it out, made it vast so people can live on it. Allah balanced its conditions and filled it with everything humans need to live: food, water, shelter, and beauty. There are signs on earth that point to the Creator, His power, wisdom, and specification of the creation. The earth is spread out like a carpet for what is above it, and it contains pathways and passages for those who move around it. It is divided into plains, mountains, hard and soft soil, fertile and saline lands that barely grow anything. There are gushing springs, varied minerals, and scattered animals of different shapes and forms.

Science of Earth

Perfect Conditions:

Earth is the right distance from the sun—not too hot, not too cold.

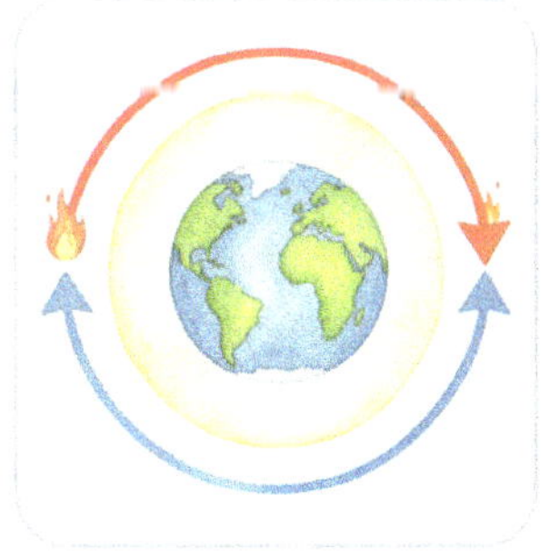

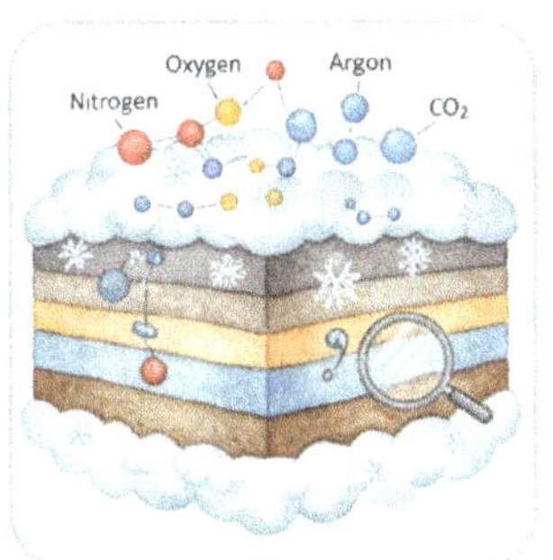

Air to Breathe:

Its atmosphere has the exact balance of gases needed for life.

Water Everywhere:

Oceans, rivers, and lakes cover most of its surface, making life possible.

Soil and Growth:

Its crust holds minerals and fertile soil for crops.

Balance and Stability:

The Earth holds us firmly and affects seasons.

Wisdom from the Scholars

Al-Qurtubi explained that the word qarār (settlement) in this verse means the earth is made livable and stable: not shaking too violently and spread out with places to dwell, farm, and travel. It is a sign that Allah did not create randomly, but with wisdom.

Lessons from the Earth

1 Prepared for Us: The earth was created as a home for humanity, not by accident.

2 Balance and Precision: Life depends on countless factors being exactly right.

3 Gratitude: Every bite of food, every breath, every shelter comes from this earth Allah spread out.

4 Reminder of Responsibility: The earth is a gift and a trust. How we treat it reflects our faith.

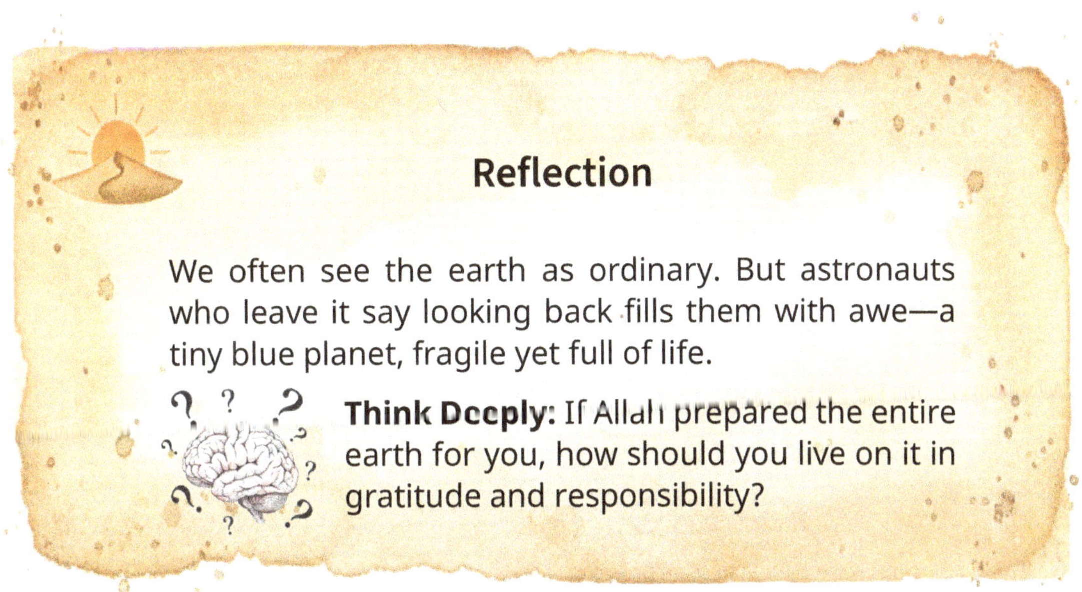

Reflection

We often see the earth as ordinary. But astronauts who leave it say looking back fills them with awe—a tiny blue planet, fragile yet full of life.

Think Deeply: If Allah prepared the entire earth for you, how should you live on it in gratitude and responsibility?

The earth is more than land and soil. It is a perfectly designed home, a stage for our test, and a constant reminder of Allah's mercy and wisdom.

> *O Allah, make us grateful for the earth You spread out for us, and let us walk upon it with humility and care.*

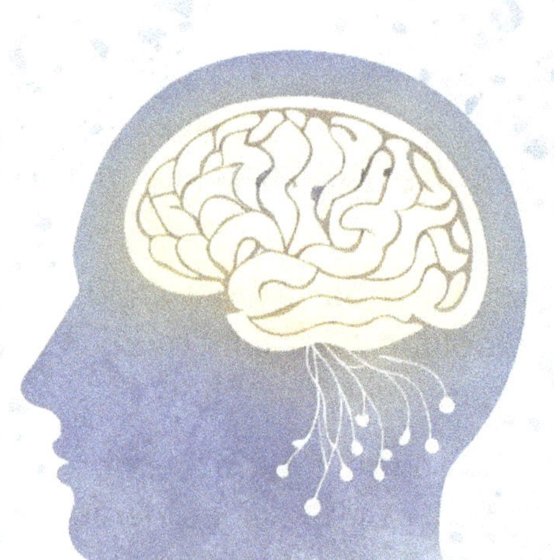

PART 4
THE MIND'S PROOFS

﴿ قُلِ انْظُرُوا مَاذَا فِي السَّمَاوَاتِ وَالْأَرْضِ وَمَا تُغْنِي الْآيَاتُ وَالنُّذُرُ عَنْ قَوْمٍ لَا يُؤْمِنُونَ ﴿١٠١﴾ ﴾

It means: *"Say. 'Observe what is in the heavens and the earth.' But signs and warnings do not benefit those who do not believe." (Yūnus, 10:101)*

We saw the signs in animals, in ourselves, and throughout the heavens and the earth. Now, let us turn to the mind itself to show that belief in Allah is not only natural, but necessary. Reason alone leads us to the certainty that Allah must exist. The Qur'an repeatedly invites us to think and reflect showing that intellect has important place in the religion of Allah. In Surat Yūnus, we are urged to observe and reflect, and when reflection is sincere and sound, it leads to the realization that all of these creations are created by Allah. The truth of Islam is deduced by contemplating the dominion of the heavens and the earth. Everything created by Allah points to His existence, knowledge, power, and will. For without an existent, living, knowing, powerful, and willing Creator, the existence of this world would not be logically possible.

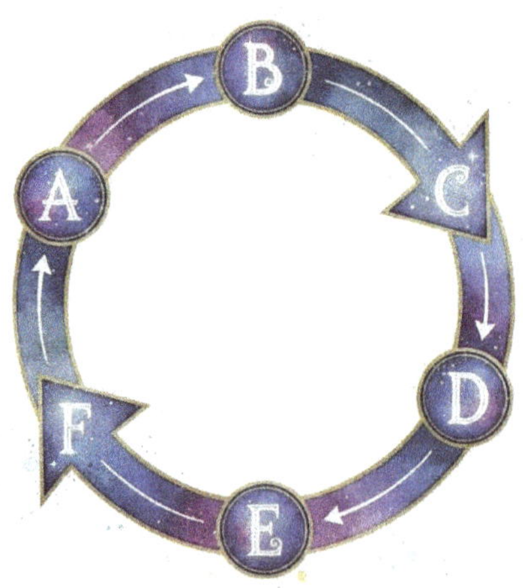

CHAPTER 19

THE IMPOSSIBILITY OF CIRCULAR CAUSATION (DAWR)

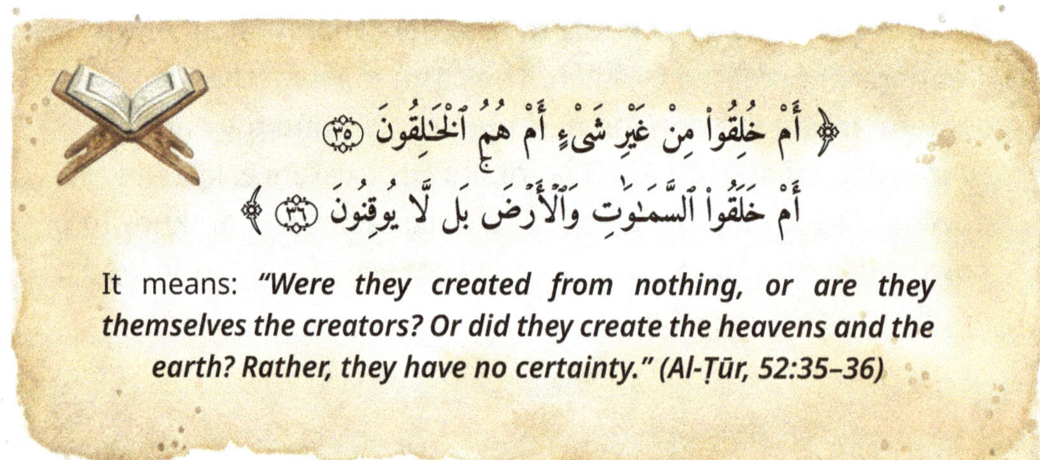

أَمْ خُلِقُوا۟ مِنْ غَيْرِ شَىْءٍ أَمْ هُمُ ٱلْخَٰلِقُونَ ٣٥ أَمْ خَلَقُوا۟ ٱلسَّمَٰوَٰتِ وَٱلْأَرْضَ بَل لَّا يُوقِنُونَ ٣٦

It means: *"Were they created from nothing, or are they themselves the creators? Or did they create the heavens and the earth? Rather, they have no certainty."* (Al-Ṭūr, 52:35–36)

These verses provide you with logical undeniable proof:

1 You were not created without a creator.

2 You did not create yourself.

<table>
<tr><td>

3 Nothing creates itself.

</td><td>

4 Humans did not create the universe.

</td></tr>
<tr><td>

5 Therefore, a Creator must exist.

</td><td>

6 Those who still deny do so out of uncertainty, not truth.

</td></tr>
</table>

The Qur'an teaches us that disbelief is not built on certainty—true certainty belongs to those who recognize the Creator.

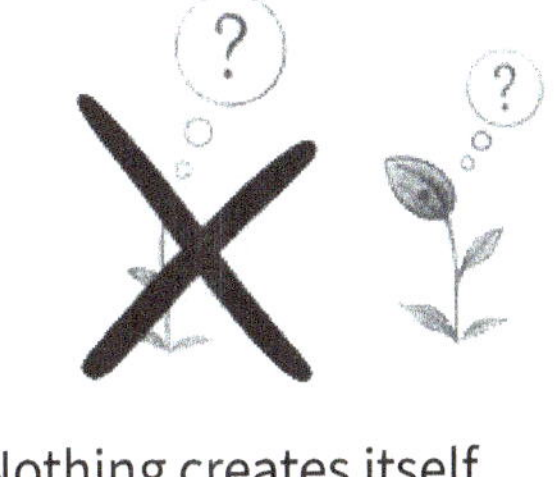

Nothing creates itself

Therefore, a **Creator** *must exist*

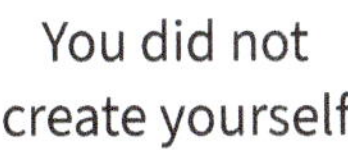

You did not create yourself

Humans did not create the Universe

Those who still deny do so out of uncertainty, not truth

Can Something Create Itself?

If someone claims there is no Creator, then they must explain how things came into existence. One possibility, they might say, is that something created itself. But that makes no sense. How can something exist before it exists in order to bring itself into existence?
It would be like saying: "This book wrote itself." For a book to write itself, it would already need to exist before existing which is a contradiction.

Circular Causation (Dawr)

Scholars call this idea dawr or circular causation. Imagine someone asks:

*   "Who built this house?"
*   One answers: "The house built itself."

This is impossible, because building requires a builder who exists before the house. In the same way, the universe could not have brought itself into existence.

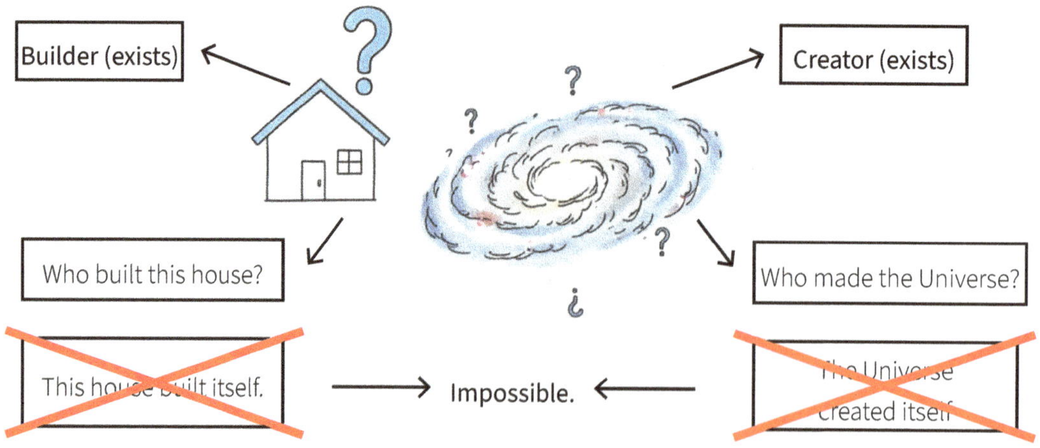

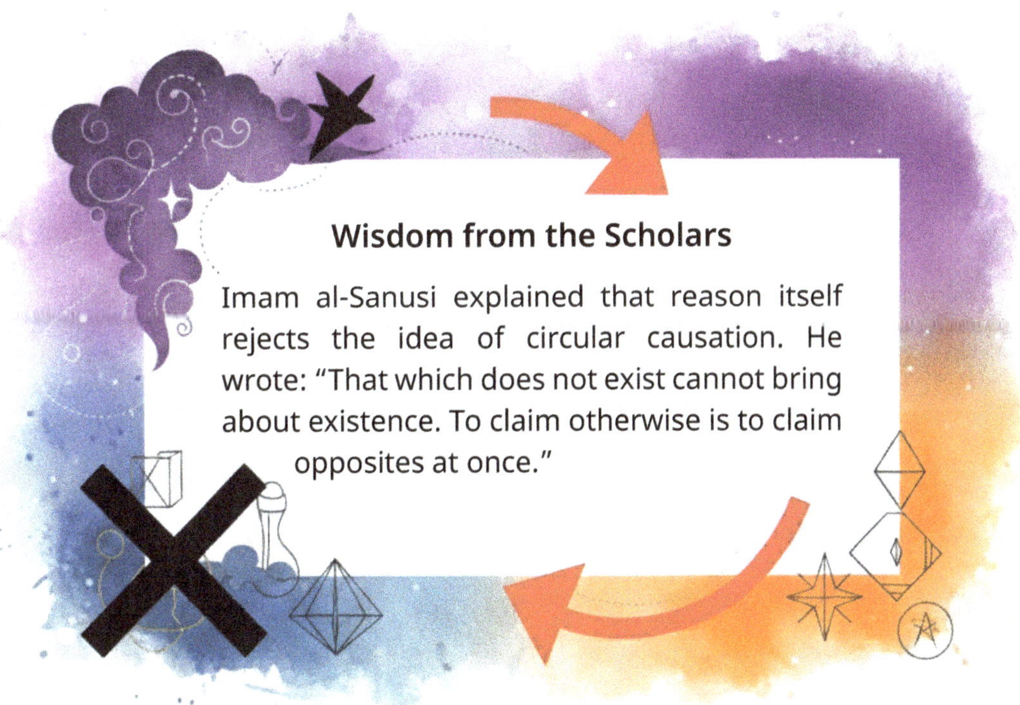

Lessons from the Dawr

1   **Contradiction:** To create itself, a thing must exist and not exist at the same time which is impossible.

2   **Reason and Revelation Agree:** The Qur'an exposes this false idea directly: "Were they created from nothing, or are they themselves the creators?"

3   **Dependence on Allah:** Since nothing can create itself, everything must be created by a Creator who existed before all else.

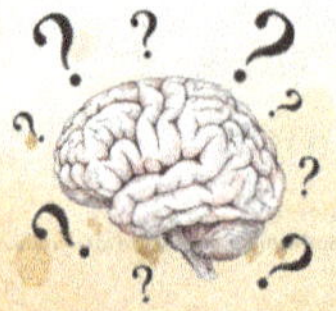

Reflection

Think of your phone. Could it create itself? Could the screen, battery, and circuits decide to form out of nothing and assemble themselves? If that's impossible for a phone, how much more for the universe?

Think Deeply: If you reject Allah as the Creator, what alternative makes sense—that creation came from nothing, or that it created itself? Both are impossible.

Circular causation is impossible. Nothing can create itself. The only logical answer is that Allah, who has always existed, created everything.

> **O Allah, You are the First, and there is nothing before You. Grant us certainty in Your existence and power.**

This universe needs a Creator.

Every building needs a builder.

Every letter needs a writer.

CHAPTER 20

THE IMPOSSIBILITY OF INFINITE REGRESS (TASALSUL)

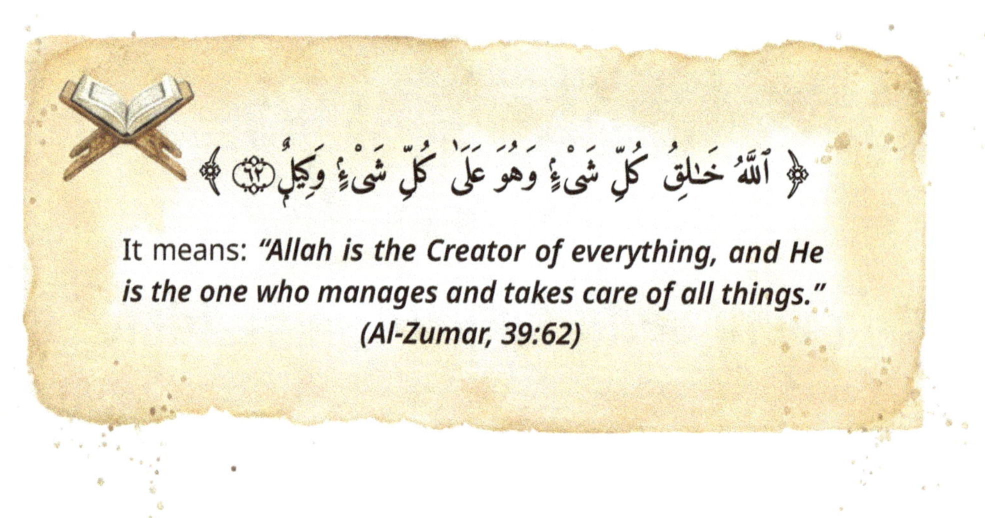

A Chain That Never Begins?

Another claim some people make is that the universe came about through an endless chain of causes. In other words, one thing created another, which created another, and so on, stretching backwards forever.

But if there is no beginning to the chain, then nothing would exist now.

The Domino Example

Imagine a line of dominoes. Each one falls because the one before it fell. But if there was no first domino to be pushed, the chain could never begin. You would wait forever, and nothing would ever happen.

In the same way, the chain of causes in the universe cannot go back forever with no beginning. There must be a First One that started everything.

Lessons from the Tasalsul

1 **A Beginning is Necessary:** Without a first one that brought the creation into existence, nothing could exist.

2 **Allah is the First:** He has no beginning, unlike creation.

3 **Reason and Revelation Agree:** The Qur'an says plainly: "Allah is the Creator of everything."

4 **Clarity Against Doubt:** Infinite regress may appear to some clever, but it collapses under logic.

Reflection

All created things point back to Allah, the Necessary Being who began all existence and who Himself was never begun.

Think Deeply: If everything you see depends on something before it, doesn't reason demand a First One that depends on nothing?

Infinite regress is impossible. A beginningless chain of events would never lead to anything. The only answer is that Allah is the First, eternal, not in need of anything, and the Creator of everything.

> *O Allah, You are al-Awwal, the First without beginning. Let our faith in You be firm without any doubt.*

CHAPTER 21

THE ARGUMENT FROM DESIGN AND ORDER

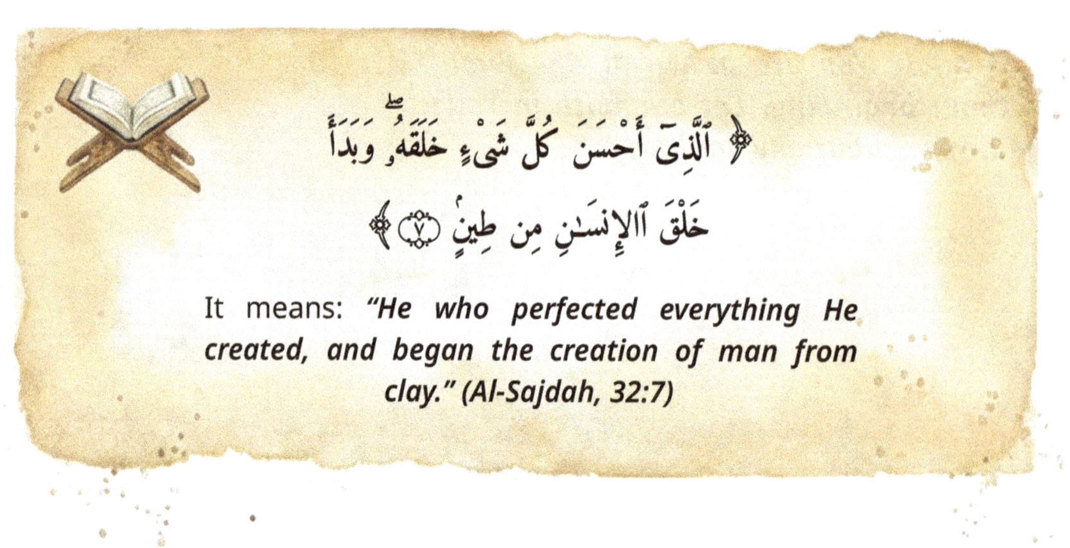

It means: *"He who perfected everything He created, and began the creation of man from clay." (Al-Sajdah, 32:7)*

Design Everywhere

Look around you: the veins of a leaf, the wings of a bird, the orbits of planets. Everything has order, structure, and purpose.

If you found a phone in the sand, would you think it formed by itself? Its circuits, screen, and apps show design. If that is true for a phone, how much more for the universe, which is tremendously more complex?

The Order of the Universe

Precise Laws:

Physics, and chemistry operate with perfect consistency.

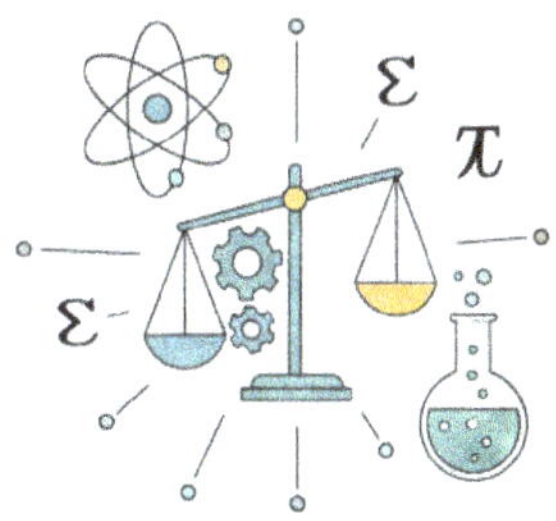

Delicate Balance:

Earth's distance from the sun and atmosphere are all determined for life.

Purposeful Function:

The eye is for seeing, the ear for hearing, the heart for pumping blood.

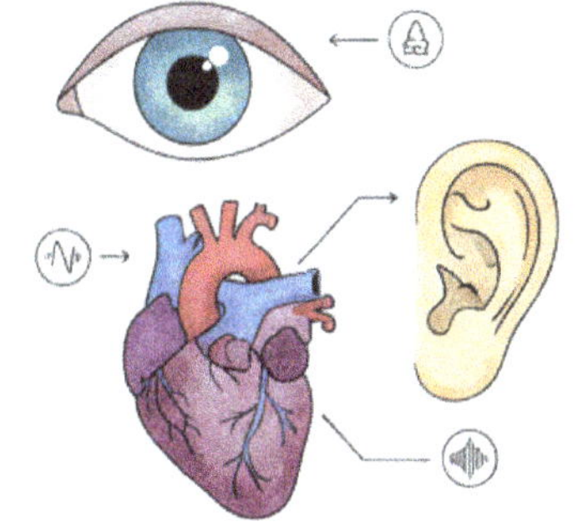

Beauty and Symmetry:

From snowflakes to galaxies, patterns repeat across creation

Order does not come from chaos by accident—it comes from will and wisdom.

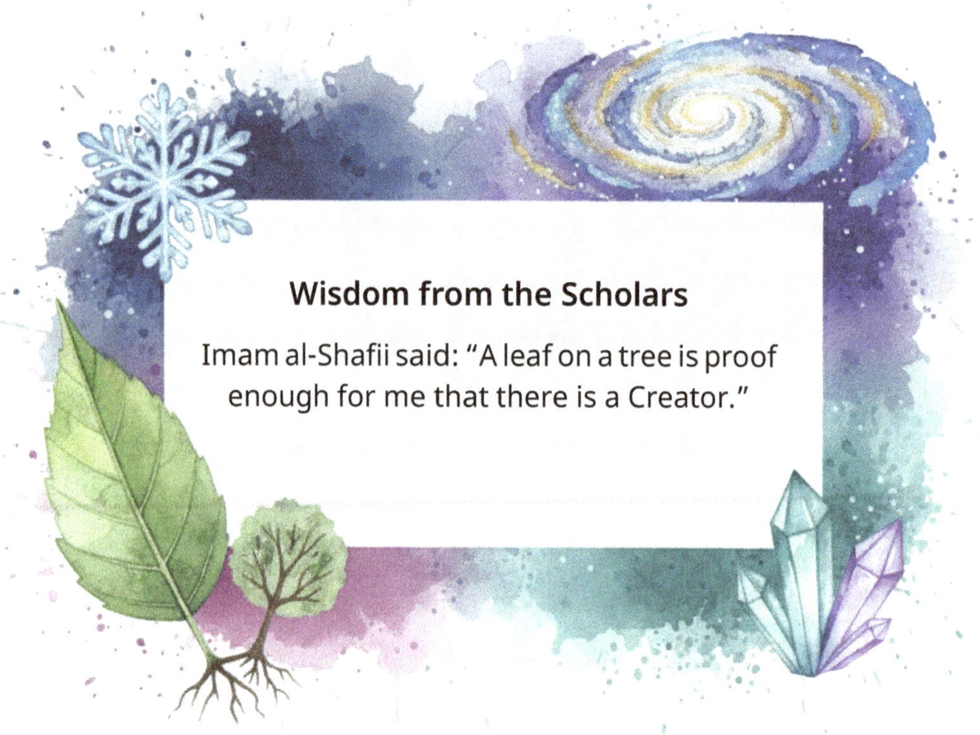

Lessons from the Design and Order

1 **Accident or Intention?** Order points to a Willing Being, not to chance.

2 **Perfection in Creation:** Allah "perfected everything He created."

3 **From Small to Vast:** The same perfection of design appears in cells and in galaxies.

4 **Gratitude and Awe:** Recognizing design should lead to worship, not neglect.

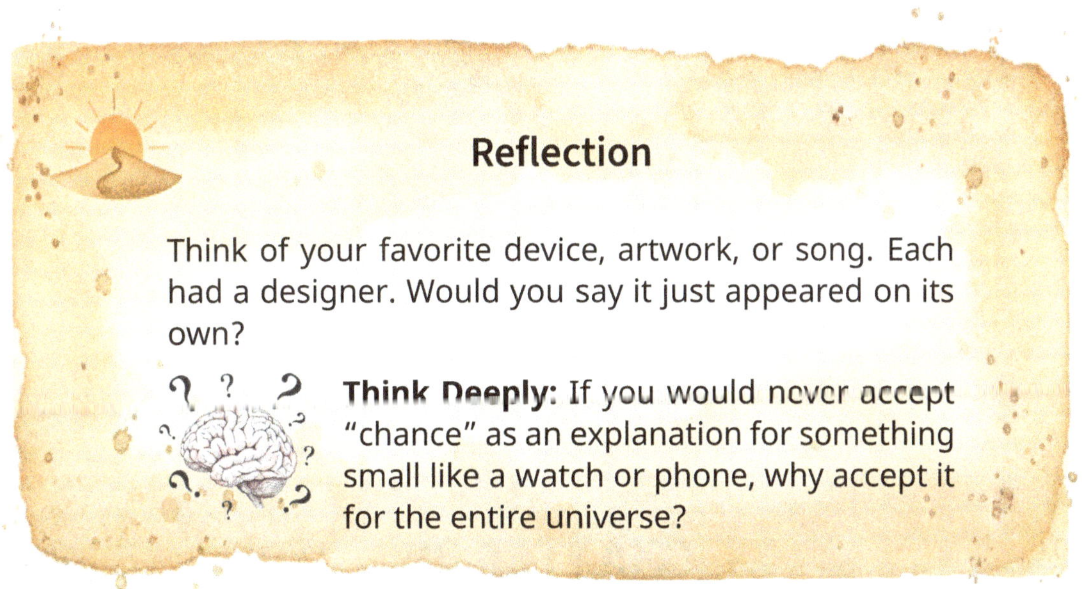

Reflection

Think of your favorite device, artwork, or song. Each had a designer. Would you say it just appeared on its own?

Think Deeply: If you would never accept "chance" as an explanation for something small like a watch or phone, why accept it for the entire universe?

The universe is full of order, precision, and beauty. Reason and revelation agree: this cannot be from chaos or chance. It is the creation of Allah, the All-Wise.

" O Allah, let the beauty and order of creation increase us in certainty and gratitude to You. "

CHAPTER 22

THE QUR'AN: A CONTINUOUS PROOF

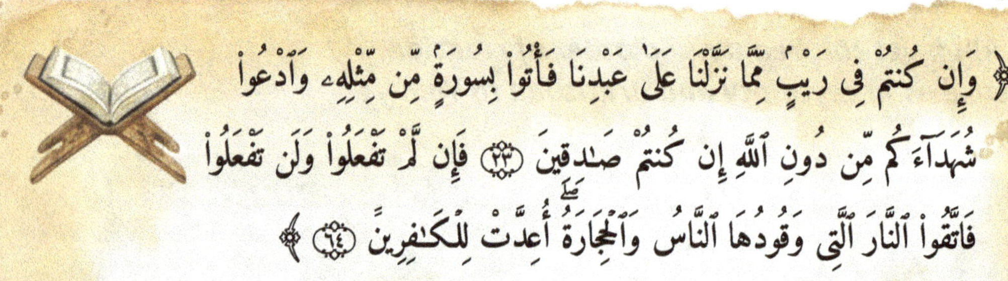

It means: *"And if you are in doubt about what We have sent down upon Our slave, then produce a surah like it, and call your witnesses besides Allah, if you are truthful. But if you do not—and you will never be able—then fear the Fire, whose fuel is people and stones, prepared for the disbelievers." (Al-Baqarah, 2:23–24)*

The Qur'an's Challenge

The Qur'an is not only a book of guidance; it is a proof in itself. It challenged the Arabs, masters of eloquence, to produce even one chapter like it. More than 1,400 years have passed, and no one has met this challenge.

Signs of the Qur'an's Divine Origin

Unmatched Eloquence:

Its style, rhythm, and depth moved even its enemies to silence.

Preservation:

Unlike other scriptures, the Qur'an has never been altered. It is memorized by millions word for word, letter for letter.

Knowledge of the Unseen:

It foretold the victory of the Romans after defeat (Surah al-Rūm, 30:2–4), which came true exactly as revealed.

Consistency:

Revealed over 23 years in different situations, yet it flows as one perfect message without contradiction

Guidance and Impact:

It transformed a society of tribes into a civilization of faith, knowledge, and justice.

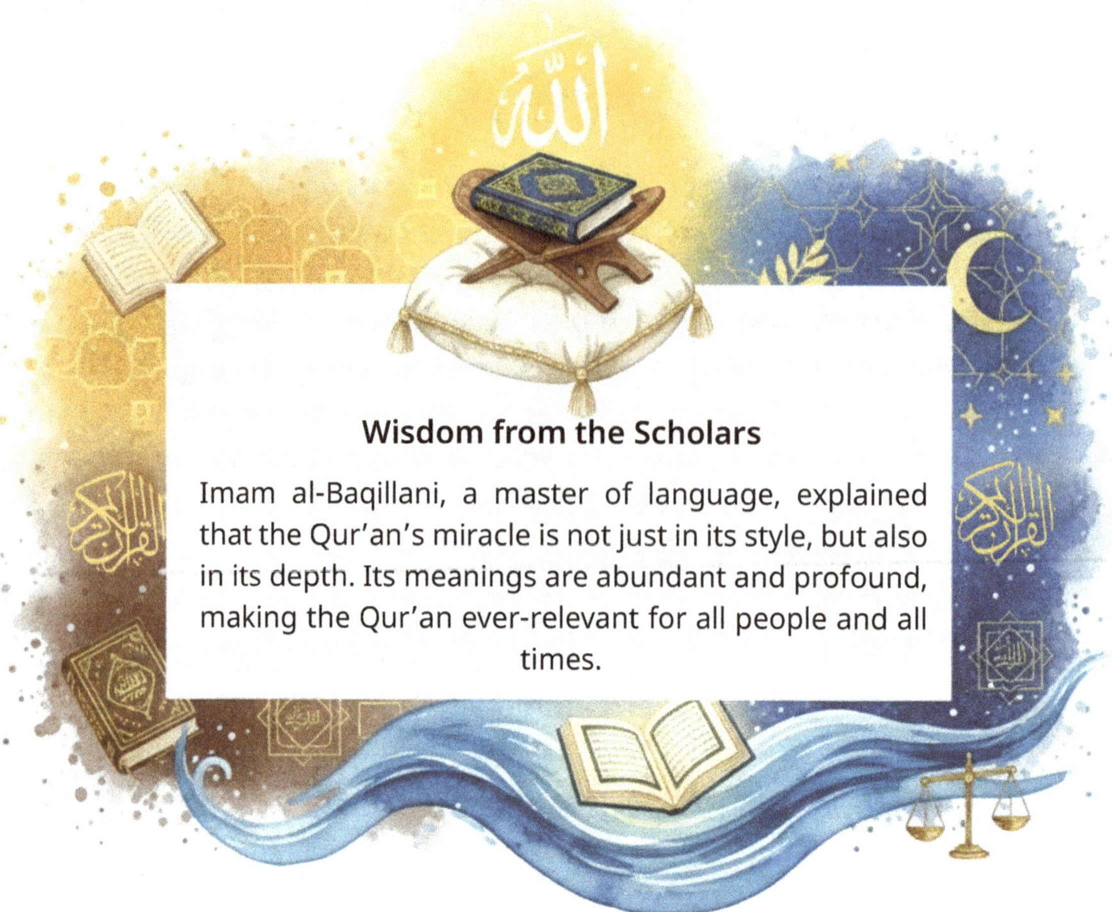

Lessons from the Qur'an

1   **A Living Miracle:** Unlike past miracles limited to their time, the Qur'an is accessible to every generation.

2   **Proof Beyond Doubt:** Its eloquence, preservation, and fulfilled prophecies testify to its divine source.

3   **Guidance for Life:** It is not only a proof, but a guidance for how to live.

4   **Responsibility:** Since the proof is clear, ignoring it is turning away from truth.

Reflection

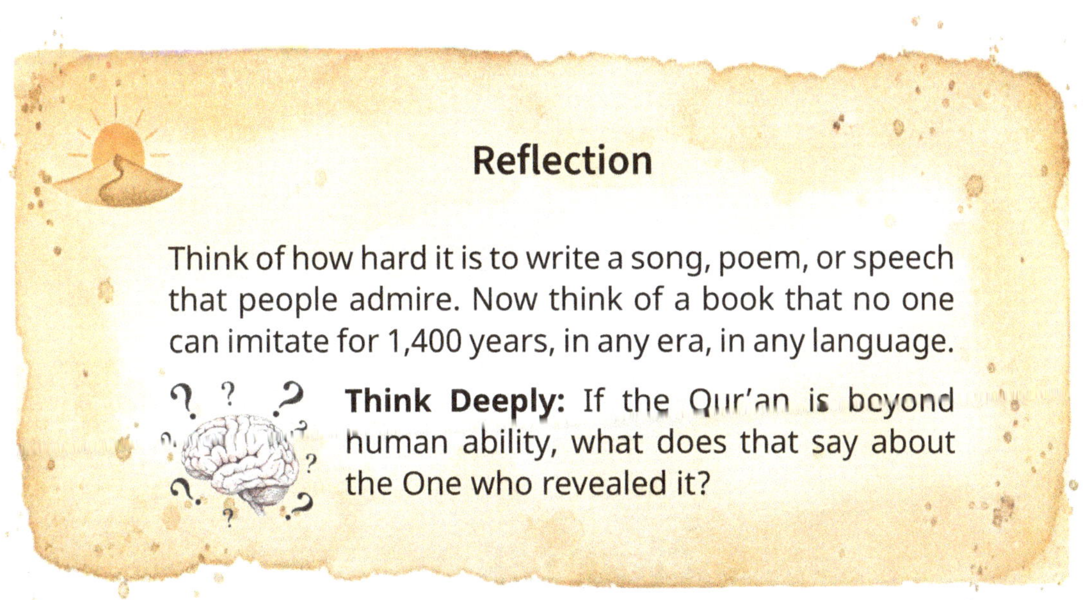

Think of how hard it is to write a song, poem, or speech that people admire. Now think of a book that no one can imitate for 1,400 years, in any era, in any language.

Think Deeply: If the Qur'an is beyond human ability, what does that say about the One who revealed it?

The Qur'an is more than words on a page. It is a miracle that speaks across time, a living proof that the message of Islam is from Allah.

❝ *O Allah, make the Qur'an the light of our hearts, the proof of our faith, and the guide of our lives.* **❞**

CHAPTER 23

THE FIṬRAH
THE NATURAL PREDISPOSITION TO ISLAM

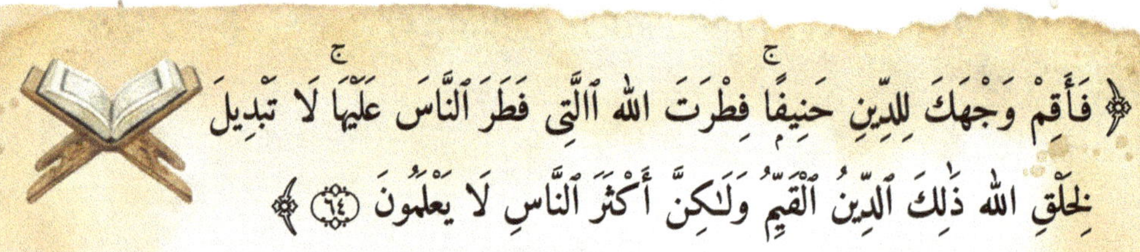

It means: *"So set your face towards the religion, inclining to truth; the fiṭrah of Allah upon which He has created people. There is no change in Allah's creation. That is the straight religion, but most people do not know." (Al-Rūm, 30:30)*

Born with Fiṭrah

Every human is born with a natural inclination: the fiṭrah. It is the inner predisposition to accept Islam since all souls acknowledged the Lordship of their Creator before the bodies were created. Children, without being taught, ask: "Who made the sky? Who made me?" The readiness to accept the belief in the Creator is instilled in us by Allah.

Signs of the Fiṭrah

Childlike Certainty:

Before doubt or philosophy, children recognize design and ask about a Maker.

Conscience:

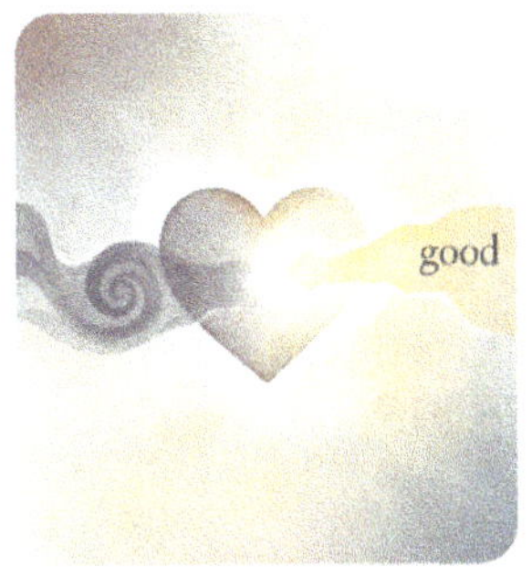

The pull toward good and discomfort with evil are part of this inner wiring.

Persistence:

Even in times of distress, many turn instinctively to God, saying "O God, help me," even if they denied Him before.

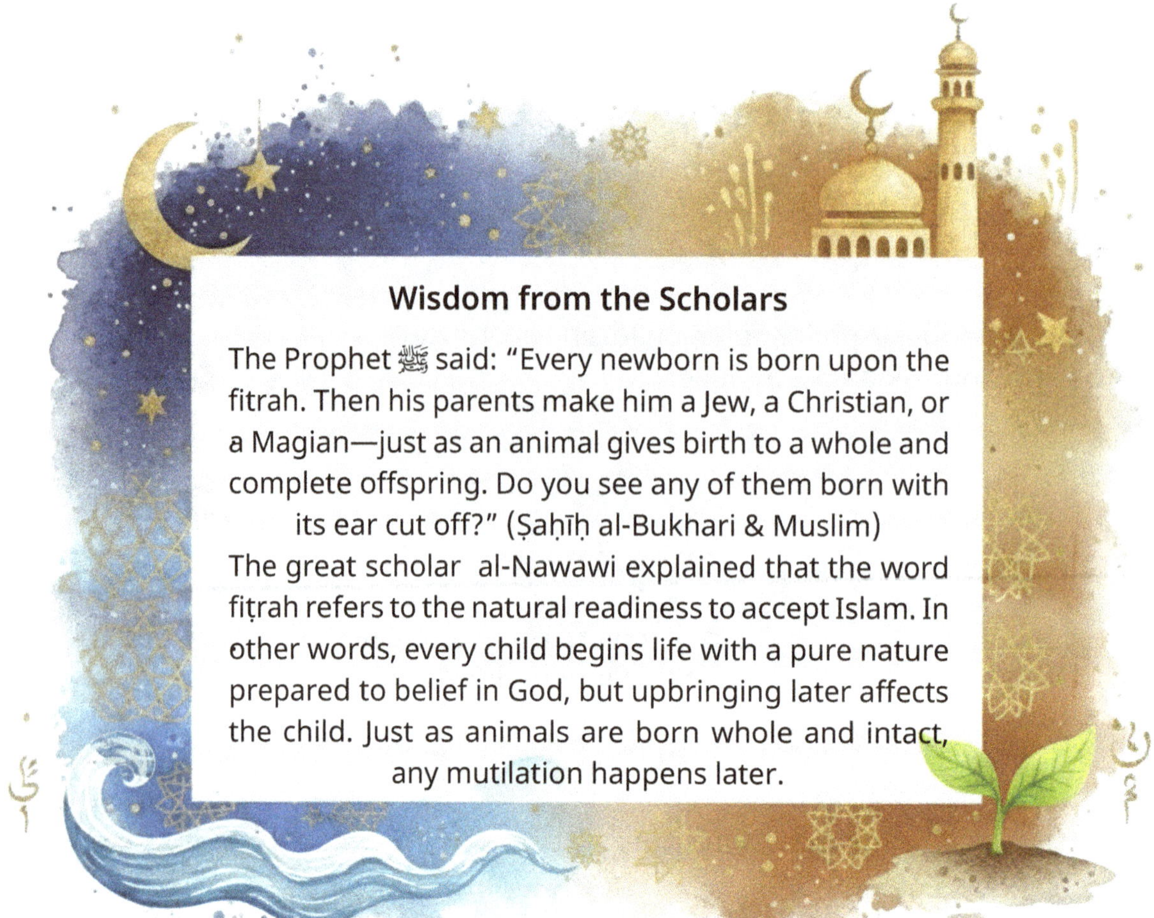

Lessons from the Fiṭrah

1   **Faith is Natural:** Atheism is not the default; belief is. Every human being is created with an innate disposition toward recognizing the Creator.

2   **Nurturing Matters:** Families and societies shape how the fiṭrah develops. The beliefs, ideas and ifluences surrounding a child as they grow can nurture or alter that original state.

3   **Return to Purity:** Islam is a return to the fiṭrah; the straight path humans are called for.

4   **Proof in Human Nature:** Our very instincts are signs that Allah is our Creator.

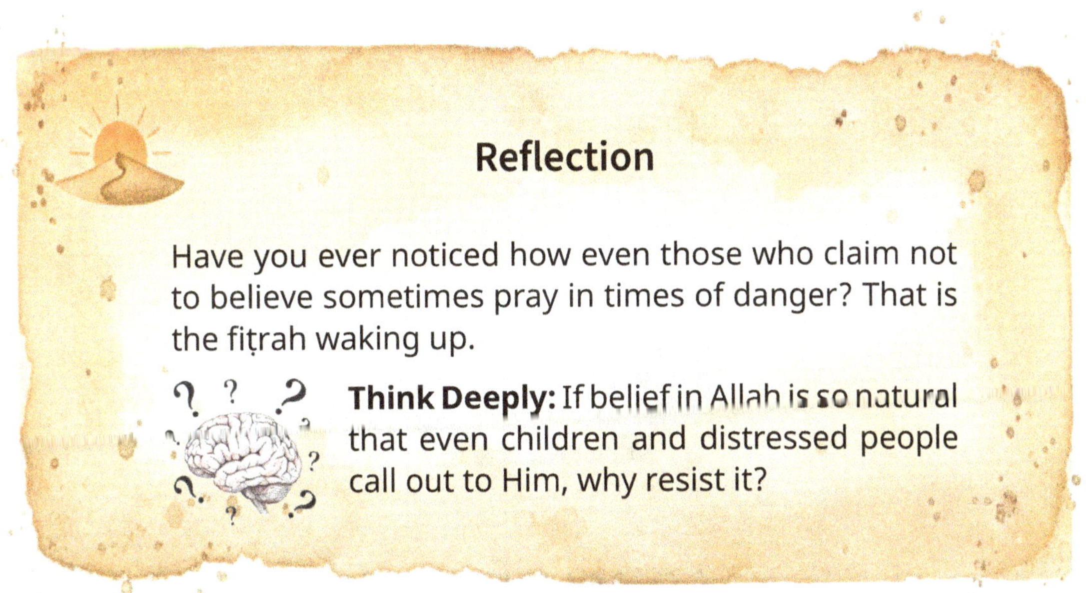

Reflection

Have you ever noticed how even those who claim not to believe sometimes pray in times of danger? That is the fiṭrah waking up.

Think Deeply: If belief in Allah is so natural that even children and distressed people call out to Him, why resist it?

The fiṭrah is the whisper of the soul to return to the state of belief it once held when it took the oath long before joining the body. It is a proof within us that no denial can erase.

" *O Allah, keep our fiṭrah pure, and let us live and die upon the truth You placed in our* **"**

CONCLUSION

It means: *"We will show them Our signs in the horizons and within themselves until it becomes clear to them that it is the Truth. But is it not sufficient that your Lord is a Witness over all things?" (Surah Fuṣṣilat, 41:53)*

The Journey You Took

You have walked through a journey of signs. Wherever you looked: in the living world around you, within yourself, in the vastness of the heavens and earth, and in the clarity of the sound mind itself, signs are present. Each perspective leads to the same truth. There must be a Creator, and He is Allah.

What the Signs Point To

Recognizing that Allah exists is only the beginning. The signs point to who this Creator is:

He is

- the First without beginning and the Last without end.
- the Creator of every detail and the Preserver of every moment.
- the One who placed predisposition to faith in your fiṭrah, proofs in your reason, and guidance in His revelation.

What creation shows, the mind confirms, and a sound heart recognizes, is affirmed by revelation and points to the same truth, leaving no room for doubt even when atheism or skepticism whisper.

A Call to Teens

You live in a time when doubts are loud and distractions are everywhere. Yet truth shines brighter than noise. Look at creation, reflect on yourself, think deeply, and you will find that faith in the existence of the Creator is light upon light. To know that Allah exists is the beginning of a life of worship!

ﷲ (**Allah**) is the name of the Creator in Arabic, **God** in English, **Бог** (**Bog**) in Russian, **Dieu** in French, خدا (**Khuda**) in Urdu, **Tanrı** in Turkish. In different languages, the name differs but the meaning remains the same: *The Creator and the Lord of all the Universe.*

See. Think. And believe with certainty.

FAITH TOOLBOX:
WHY WE KNOW ALLAH EXISTS

1. The Creator's Existence is Necessary

Everything we see is created—cells, stars, galaxies, time, space, and laws of nature.
Anything that begins to exist needs a Creator.
The universe is not eternal; therefore, it must have been created.

2. The Creator Is Eternal and Without Need

The Creator cannot Himself be created; otherwise, He would not be the Creator.
He must be:

- Eternal—has no beginning
- Free of all dependence—needs nothing
- All-Powerful—able to create from nothing
- Unlike creation—not bound by time, place, or form

3. The Signs in the Universe Point to Allah's existence

Wherever we look, we see order and precision:
- the balance of ecosystems
- the laws of physics
- the orbit of the sun and moon
- the remarkable precision of the universe

These signs are not random. They are clear proof of the wisdom of the Creator.

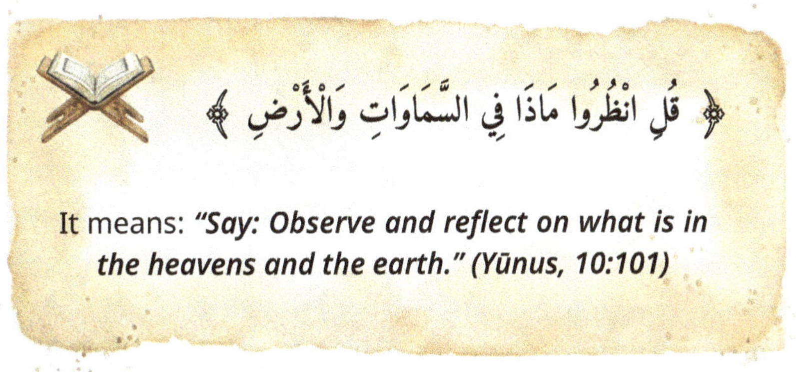

It means: *"Say: Observe and reflect on what is in the heavens and the earth."* (Yūnus, 10:101)

4. The Signs Within Ourselves

Every human being carries signs of Allah's power and wisdom:
- the mind that can reason
- the heart that feels
- the precise function of organs
- the natural inclination toward belief in a Creator

These inner signs affirm that we were not made by accident.

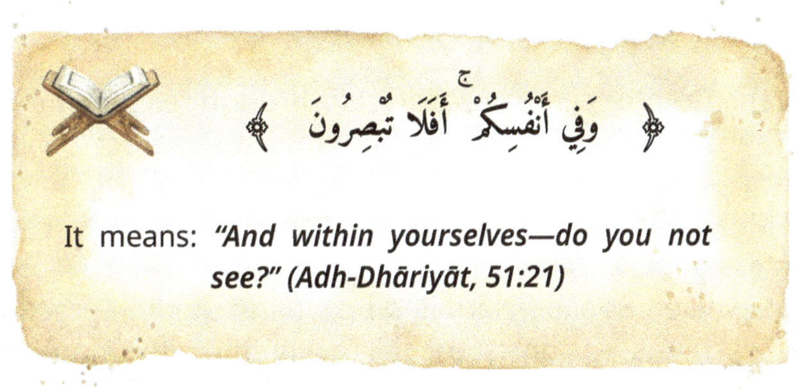

It means: *"And within yourselves—do you not see?"* (Adh-Dhāriyāt, 51:21)

5. It Is Impossible for the Universe to Create Itself

A thing cannot create itself; what doesn't exist cannot bring itself into existence.
This applies to everything in the universe including:
- matter
- energy
- time
- space

Therefore, the Creator must exist before everything else.

6. Faith Is Supported by Understanding and Reflection

Islam encourages us to think, observe, and reflect.
Faith follows a clear path supported by reason and sound understanding.

7. The Prophets Bring the Final Proof

Even though the Creator can be recognized through reason and reflection alone, Allah sent prophets to guide humanity to correct belief in Him through:
- Miracles—clear signs beyond human ability, transmitted across generations
- Revealed Books—clarifying belief and guiding practice
- Teachings—a consistent call to worship Allah alone

Through the Prophets, Allah completed the proof upon humanity.

THE PROOF FROM CHANGE

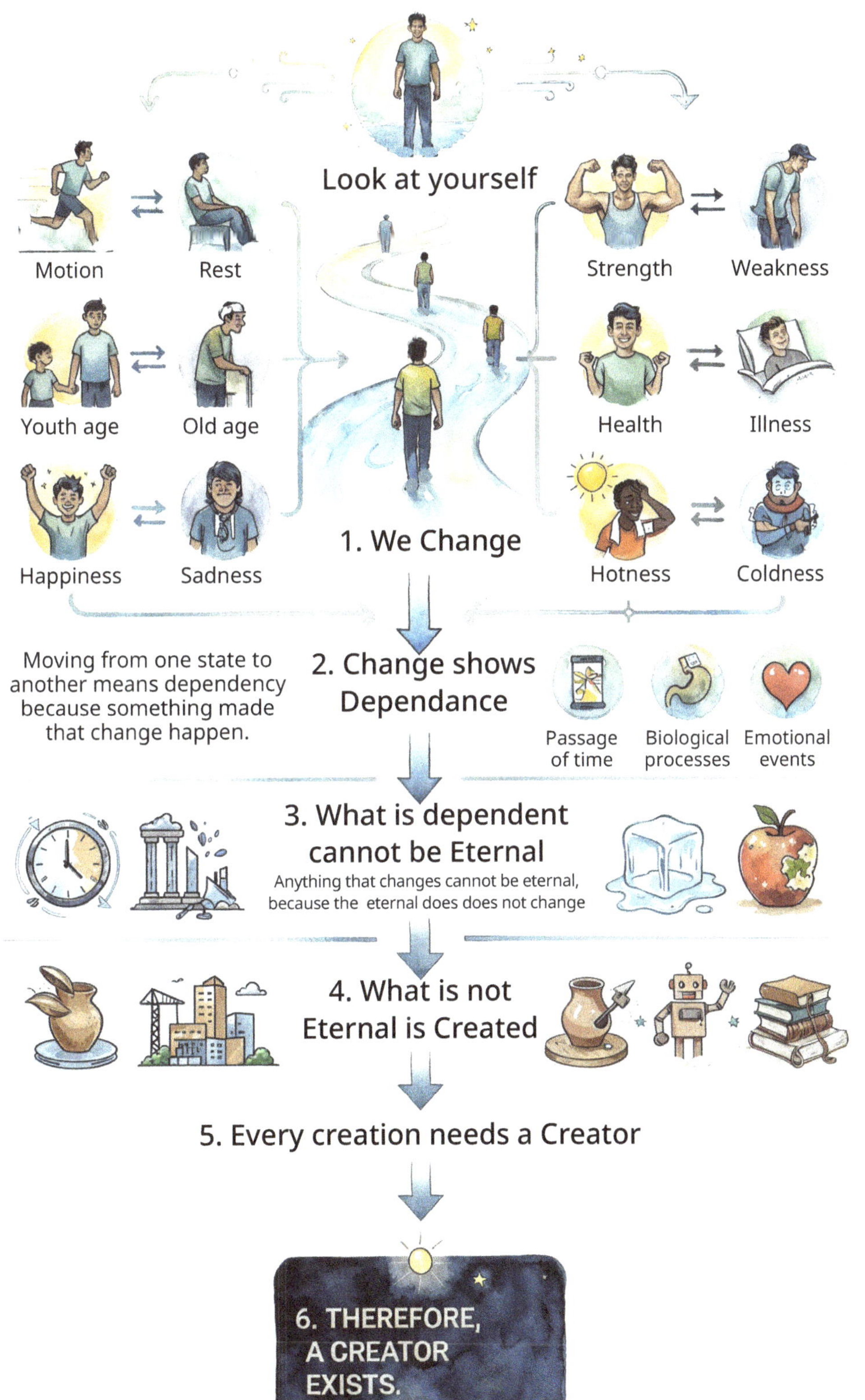